Proceedings of the International Symposium on

DIRECT ROLLING AND HOT CHARGING OF STRAND CAST BILLETS

Pergamon Titles of Related Interest

Bee MATERIALS ENGINEERING
Davenport FLASH SMELTING
Foesco FOUNDRYMAN'S HANDBOOK 9th EDITION
Reid METAL FORMING & IMPACT MECHANICS

Other CIM Proceedings
Published by Pergamon

Bickert REDUCTION AND CASTING OF ALUMINUM
Kachaniwsky IMPACT OF OXYGEN ON THE PRODUCTIVITY OF
   NON-FERROUS METALLURGICAL PROCESSES
Macmillan QUALITY AND PROCESS CONTROL IN REDUCTION AND
   CASTING OF ALUMINUM AND OTHER LIGHT METALS
Plumpton PRODUCTION AND PROCESSING OF FINE PARTICLES
Rigaud ADVANCES IN REFRACTORIES FOR THE
   METALLURGICAL INDUSTRIES
Ruddle ACCELERATED COOLING OF ROLLED STEEL
Salter GOLD METALLURGY
Thompson COMPUTER SOFTWARE IN CHEMICAL AND
   EXTRACTIVE METALLURGY
Tyson FRACTURE MECHANICS
Wilkinson ADVANCED STRUCTURAL MATERIALS

Related Journals
(Free sample copies available upon request)

ACTA METALLURGICA
CANADIAN METALLURGICAL QUARTERLY
MINERALS ENGINEERING
SCRIPTA METALLURGICA

Proceedings of the International Symposium on

DIRECT ROLLING AND HOT CHARGING OF STRAND CAST BILLETS

Montréal, Canada
August 29-30, 1988

Co-sponsored by
the Canadian Steel Industry Research Association (CSIRA),
the Canadian Continuous Steel Casting Research Group (CCSRG),
and the Iron and Steel Section of the Canadian Institute
of Mining and Metallurgy

Vol. 10 Proceedings of the Metallurgical Society
of the Canadian Institute of Mining and Metallurgy

Edited by

J.J. JONAS

CSIRA-NSERC Professor of Steel Processing Department of Metallurgical Engineering,
McGill University, Montréal, Canada

R.W. PUGH

Senior Research Associate, Steltech, a subsidiary of Stelco Inc., Edmonton, Canada

S. YUE

CSIRA-NSERC Research Associate, Department of Metallurgical Engineering,
McGill University, Montréal Canada

PERGAMON PRESS

New York   Oxford   Beijing   Frankfurt   São Paulo   Sydney   Tokyo   Toronto

Pergamon Press Offices:

U.S.A.	Pergamon Press, Inc., Maxwell House, Fairview Park, Elmsford, New York 10523, U.S.A.
U.K.	Pergamon Press plc, Headington Hill Hall, Oxford OX3 0BW, England
PEOPLE'S REPUBLIC OF CHINA	Pergamon Press, Qianmen Hotel, Beijing, People's Republic of China
FEDERAL REPUBLIC OF GERMANY	Pergamon Press GmbH, Hammerweg 6, D-6242 Kronberg, Federal Republic of Germany
BRAZIL	Pergamon Editora Ltda., Rua Eça de Queiros, 346, CEP 04011, São Paulo, Brazil
AUSTRALIA	Pergamon Press (Aust.) Pty Ltd., P.O. Box 544, Potts Point, NSW 2011, Australia
JAPAN	Pergamon Press, 8th Floor, Matsuoka Central Building, 1-7-1 Nishishinjuku, Shinjuku-ku, Tokyo 160, Japan
CANADA	Pergamon Press Canada Ltd., Suite 271, 253 College Street, Toronto, Ontario M5T 1R5, Canada

First printing 1989

Library of Congress Cataloging in Publication Data

International Symposium on Direct Rolling and Hot Charging of Strand Cast Billets (1988 : Montréal, Québec)
   Proceedings of the International Symposium on Direct Rolling and Hot Charging of Strand Cast Billets, Montreal, Canada, August 29-30,
   1988 / co-sponsored by the Canadian Steel Industry Research Association (CSIRA), the Canadian Continuous Steel Casting Research Group (CCSRG), and the Iron and Steel Section of the Canadian Institute of Mining and Metallurgy ; edited by J.J. Jonas, R.W. Pugh, S. Yue.
   p.   cm. -- (Proceedings of the Metallurgical Society of the Canadian Institute of Mining and Metallurgy ; vol. 10)
   Includes bibliographies and indexes.
   ISBN (invalid) 008036998
   1. Steel founding--Congresses. 2. Continuous casting--Congresses.
3. Rolling (Metal-work)--Congresses. I. Jonas, J. J. II. Pugh, R. W. III. Yue, S. IV. Canadian Steel Industry Research Association. V. Canadian Continuous Steel Casting Research Group. VI. Metallurgical Society of CIM. Iron and Steel Section. VII. Title. VIII. Series.
TS233.I523   1988
672.2--dc19   89-3060
                                       CIP

In order to make this volume available as economically and as rapidly as possible, the authors' typescripts have been reproduced in their original forms. This method unfortunately has its typographical limitations but it is hoped that they in no way distract the reader.

Printed in the United States of America

FOREWORD

It is clear that the utilisation of the residual heat of casting in the subsequent hot rolling stage will yield significant energy savings, and, in conjunction with continuous casting, will greatly improve productivity. However, the execution of this simple concept requires a casting that is free from surface defects. The development of direct rolling and hot charging is thus geared, not only towards bypassing or at least decreasing the energy consumption of the reheating stage, but also towards the elimination of surface conditioning of the cast billet.

The industrial importance of direct rolling and hot charging is underscored by the fact that this international symposium was organized by the Canadian Steel Industry Research Association (CSIRA), a group that is comprised of twelve of the principal steel producers in Canada, in collaboration with the Canadian Continuous Steel Casting Research Group (CCSRG) and the Iron and Steel Section of the Canadian Institute of Mining and Metallurgy. It was held in Montreal on August the 29th and 30th at the 27th Annual Conference of Metallurgists.

The organizing commitee of this conference selected four main themes. Surface quality is obviously of major importance, and two sessions were allotted to cover this broad subject, one centered on *improving* surface quality through casting practice, the other concerned with the *detection* of surface flaws using sensors. A session on temperature equalization techniques and equipment was an important addition, not only for hot charging but also for direct rolling in which there are always problems associated with the uneven cooling of billets. Finally, steel producers from Europe, Japan and North America related their experiences with direct rolling and hot charging.

In addition to the undersigned, the CSIRA organizing committee included A. Krzakowski (Sidebec-Dosco), B. Bowman (Slater Steel), H. Shimizu (Manitoba Rolling Mills), J. G. Metrakos (Ivaco), R. Hadden (Lasco) and G. E. Ruddle (CANMET-MTL), while G. Kamal participated in the work of the CCSRG committee. To the above colleagues, as well as to the authors, session chairmen and others who contributed to the success of the symposium, we are truly grateful.

J. J. Jonas
R. W. Pugh
S. Yue

November 30th, 1988

Montreal, Quebec

v

TABLE OF CONTENTS

SESSION 1

CASTING PRACTICE AND BILLET QUALITY

Chairpersons: **R. Hadden (Lasco)**
                            D.A.R. Kay (McMaster University)

MOLD AND OSCILLATOR CHANGES FOR IMPROVED BILLET QUALITY AND CASTER PERFORMANCE

R. W. Pugh*, P. R. Staveley**, I. V. Samarasekera***, and J. K. Brimacombe***

* Steltech, 1375 Kerns Road, Burlington, Ontario L7P 3H8
** Stelco Steel, PO Box 2348, Edmonton, Alberta T5J 2R3
*** The Centre for Metallurgical Process Engineering, The University of British Columbia, Vancouver, BC V6T 1W5

ABSTRACT

A new mold and oscillator system was required on the Stelco Steel, Edmonton Works, caster for improved reliability, reduced breakout frequency and enhanced billet quality. Trials carried out on the original system indicated that a mechanical oscillator with sinusoidal motion for short stroke length, high oscillation rate, low negative-strip time operation was desired. Higher water velocity, improved lubricant distribution, thicker mold tubes and improved mold tolerances were also recommended. Implementation of the new system has reduced caster delays, reduced lubricant consumption and improved billet quality.

KEYWORDS

Continuous casting, steel billet, mold design, oscillator design, billet quality.

INTRODUCTION

Stelco Steel operates a minimill-style billet casting shop in Edmonton, Alberta, which consists of two 68-tonne electric arc furnaces and a three-strand, straight-mold caster.

TABLE 1 Edmonton Works Facilities

Furnaces		Caster	
Size (tonnes) Number Power (MVA)	68 2 1 x 45 1 x 60	Type Strands Billet Sizes (mm) Level Control Strand Support Tundish Cars Ladle Cars	Koppers, Straight Mold 3 110, 120, 150, 200 Square, some rectangular sizes. Automatic (radiation detection type) Footrolls at bottom of mold Two for "flying" tundish switches Two for sequence casting

With this shop configuration, output is limited by the caster operation. Therefore, any caster delay or breakout which slows the caster cycle reduces shop output. Most of the billets are rolled to bars at Edmonton but any excess billet capacity can be shipped to Hamilton for rolling.

All billets produced in this shop must meet shape specifications and about one-half the billets must meet surface critical specifications. Improved billet quality is required to increase the portion of product meeting these specifications.

Billet integrity and quality begin in the mold. A study was carried out at Edmonton Works in conjunction with the Centre for Metallurgical Process Engineering to indicate the key mold parameters affecting caster operation and product quality. Based on this work, a new mold and oscillator system was implemented at the Edmonton Works plant. The results of this study, and preliminary results after implementation of this system, are discussed in this paper.

EXISTING MOLD AND OSCILLATOR SYSTEM

The 1970 vintage Koppers caster at Edmonton Works utilized a hydraulic oscillation system (Fig. 1). A hydraulic cylinder attached to the mold table through a lever arm provided the oscillating motion. Limit switches on the oscillating arm controlled the flow of hydraulic fluid to either the top or bottom of the cylinder. The oscillation speed was synchronized with the casting speed by a link between the speed tachometer and the hydraulic fluid pump. Oscillation speed relative to billet speed was controlled by a "mold lead" potentiometer. Casting speed, as in most casters, was controlled by the pinch roll drives based on signals from a radiation-type level detection system.

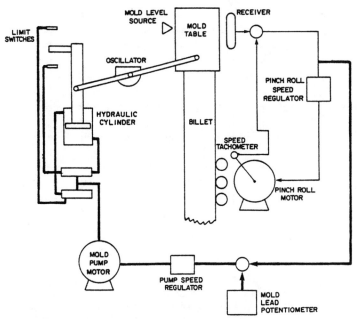

Fig. 1. Hydraulic oscillation system

The hydraulic oscillation system, especially the limit control switches, gave poor mechanical and electrical reliability. In addition, the system allowed only limited control of the oscillating motion. Oscillation frequency and stroke length were both controlled by the position of the limit switches. The hydraulic oscillator produced a sawtooth, rather than a sinusoidal, waveform (Fig. 2), which introduced high stresses during the sudden direction changes at the top and bottom of the stroke. A fairly long stroke length was required to limit stresses on the system and resulted in low oscillation rates and long negative-strip times. Typically, stroke length was set at 19 mm which gave oscillation rates of 110 cycles a minute and negative-strip times of about 0.27 s. Billets exhibited deep oscillation marks and a sensitivity to transverse cracking.

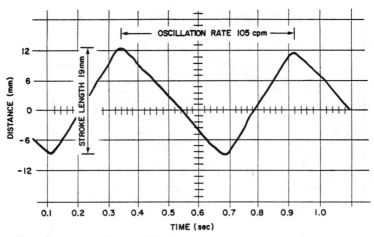

Fig. 2.   Sawtooth oscillation motion

The mold system on this caster utilized keeper plates to constrain and align the straight mold tubes. Poor tolerances resulted in a nonuniform water channel. The mold pump design and flow restrictions limited mold water velocity to about 8 m/s on the smallest mold size and 6 m/s on the largest size. An open discharge from the mold made it difficult to prevent the entry of mold lubricating oil or other contaminants into the water system. Thin mold tubes (9 mm) were susceptible to distortion, particularly on the larger billet sizes.

Billet rectangularity, the difference in billet dimensions on one side compared to the other side, is more of a problem on a straight mold caster than on a curved mold machine. On the straight mold caster, the pinch rolls are used to support the vertical weight of the strand and require sufficient pressure to prevent the strand from slipping. This situation causes a slight reduction in billet dimensions on the sides perpendicular to the pinch rolls. On the curved mold machine, the vertical weight of the strand is less and is partially supported by the guiding rolls so that less pressure is needed at the straightener to prevent slipping, resulting in less rectangularity on the curved mold machine. Changes were required to reduce rectangularity on the Edmonton Works' caster.

CASTER TRIALS

Preliminary work was carried out to monitor mold tube shape. An electronic device utilizing an LVDT* measuring head was developed by Steltech to measure internal mold tube dimensions with an accuracy of ±0.025 mm (Fig. 3). Periodic excessive mold tube distortion could be traced to water quality problems. Mold lubricant, which entered the water system through the open mold drain, would build up in the mold tank to form a latex-like material that deposited on the cooling faces of the mold tube, drastically reducing heat removal and causing extreme distortion. Improved sealing of the drain reduced these problems.

---

* Linear Variable Displacement Transducer.

 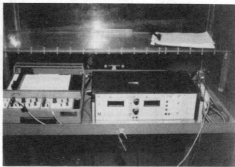

(a) Measuring head and platform   (b) XY plotter and signal conditioner

Fig. 3. Steltech tube measurement system

Ongoing monitoring of the tube dimensions continued to show limited distortion, even with improvements in the mold water system to control water hardness and iron levels. Increased mold water velocity, using two mold water pumps, failed to reduce this type of distortion. Thus, it was felt that mold design changes for improved tube support were required.

A series of trials was organized in cooperation with the Centre for Metallurgical Process Engineering at the University of British Columbia to help understand this problem and indicate further changes for improved performance. This work was carried out on the 110 mm billet size.

One mold housing was modified by remachining the water jacket and providing top, bottom and retaining plates with the correct tolerances to ensure a uniform water gap and to restrict tube movement.   A double-taper, 9 mm thick tube, instrumented with thermocouples down the centerline of all four faces, was employed in the test housing (Fig. 4). Mold water velocity was maintained at about 8 m/s. Trials were carried out at stroke lengths of 19 mm and 13 mm (negative-strip times of 0.27 and 0.23 s, respectively) using the existing, non-sinusoidal oscillator.

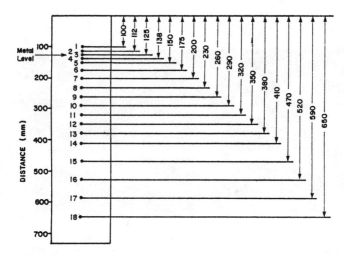

Fig. 4.   Tube thermocouple arrangement

The double taper on the tube was expected to improve heat transfer, providing a thicker and more uniform shell. At the same time, the improved attention to tolerances and mold support would ensure good water cooling. The shorter negative-strip times were expected to reduce oscillation mark depth and improve billet surface quality.

Casting conditions were monitored and billet samples were collected for three grades of steel (low, medium and high carbon) on both the test mold and a regular control mold. Mold thermocouple temperatures were monitored for each heat on the test mold. The results of the study are reported in the following sections.

Shell Thickness

The shell thickness profile was monitored by observing the location of dark bands on macroetched transverse billet sections (Fig. 5). These bands outline the thickness profile, usually just below the mold, and indicate a sudden change in heat extraction rate. Measurement of the dark band width at the off-corner and midface positions showed more uniform shell thickness on the test mold than on the control mold. Also, the uniformity was greater in the short stroke (short negative-strip time) trials than in the long stroke (long negative-strip time) trials.

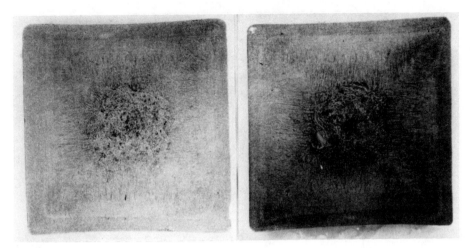

Fig. 5. Billet shell thickness profile

The improvement in uniformity of the dark bands is due to the better constraint and greater upper taper on the test mold compared to the control mold.

Oscillation Mark Depth

Oscillation mark depth was monitored on billet samples taken during the trials using a profilometer described by Bakshi and coworkers[1] (1988). Oscillation mark depth was greatest on the low-carbon billets, as expected. No significant difference could be observed between billets from the test mold and those from the control mold. However, oscillation mark depth decreased with decreasing stroke length and corresponding lower negative-strip time for both molds (Fig. 6), particularly on the low-carbon grades.

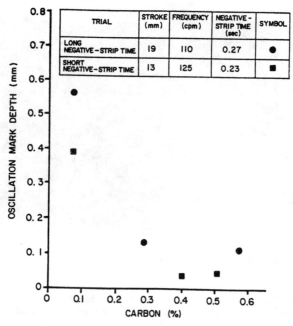

TRIAL	STROKE (mm)	FREQUENCY (cpm)	NEGATIVE-STRIP TIME (sec)	SYMBOL
LONG NEGATIVE-STRIP TIME	19	110	0.27	●
SHORT NEGATIVE-STRIP TIME	13	125	0.23	■

Fig. 6. Oscillation mark depth versus carbon content

In some instances, transverse cracks could be seen associated with the deeper oscillation marks on low-carbon grades (Fig. 7). Subsequent castings of low- carbon steels carried out at the shorter stroke length (13 mm versus 19 mm) and correspondingly shorter negative strip-time (0.23 s versus 0.27 s) with regular molds confirmed that the reduced oscillation mark depth was resulting in a reduced frequency of transverse cracks (Fig. 8). Billet samples with no cracks increased from 56 to 81 percent, while samples with cracks of depth greater than 1 mm decreased from 20 to 6 percent and samples with cracks shallower than 1 mm decreased from 29 to 13 percent.

Fig. 7. Transverse crack
associated with
deep oscillation
mark

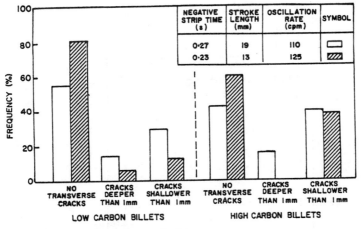

Fig. 8. Transverse crack frequency

Although the oscillation mark depth reduction was less on the high- carbon grades than on the low-carbon grades, casting with the shorter stroke and shorter negative strip times on these grades resulted in a reduced frequency of transverse cracks as well. Billet samples with no cracks increased from 43 to 61 percent while samples with cracks greater than 1 mm decreased from 16 to 0 percent and samples with cracks shallower than 1 mm remained at about 39 percent.

Heat Transfer

The time-averaged tube temperatures for selected heats on the trial mold were converted to hot-face heat-flux profiles with the aid of a mathematical model (Samarasekera and Brimacombe[2,3]). As found previously, the heat transfer is lowest for steels with low carbon content. For steels with medium to high carbon content, the heat-flux profiles are higher and almost identical. Heat transfer tends to be governed by the shell-to-mold air gap established in part by the oscillation mark depth. The lower heat transfer with the low-carbon grades is due to the deeper oscillation marks.

The peak heat-flux on the low-carbon grades was higher during the short stroke trial than during the long stroke trial (4000 kW/m^2 versus 3000 kW/m^2), which is consistent with the shallower oscillation marks seen on this grade during the short stroke run. On the medium and high-carbon grades, the peak heat-flux was also higher in the short stroke trial than that in the long stroke trial (5600 kW/m^2 versus 4800 kW/m^2), even although no measurable difference in oscillation mark depth was observed.

Furthermore, differences in heat flux were noted from face to face in the tube. These differences were considered to be related to inconsistent mold lubrication, which pointed out the need for an improved mold lubricant distribution system.

Although no temperature measurements were made on the single-taper tube used as the control mold during this trial, comparison of the double-taper data to single-taper data from trials in another plant revealed significant differences (Fig. 9).

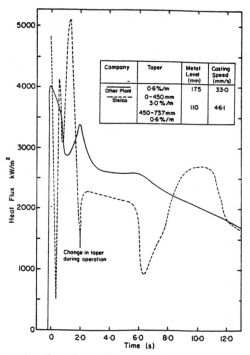

Fig. 9.   Heat flux profile

Heat transfer near the meniscus (0 to 2.0 s) on the double-taper tube shows sharp increases and decreases. Also heat transfer below the meniscus decreases and then increases at 6 s (about 400 mm from the top of the tube). Heat transfer on the single-taper tube shows a fairly consistent drop from the meniscus to the exit.

The nonuniform heat transfer from top to bottom of the double-taper tube occurs because the tube taper does not match the shrinkage that is taking place in the solidifying shell. This situation gives rise to alternate regions of contact and noncontact as shown by the increase and decrease in the heat-transfer rate. Measurement of the tube profile after the trials revealed that contact below the meniscus and at the 400 mm level had resulted in a tube dimension change. More rigid tubes and an improved taper were required.

RECOMMENDATIONS FOR REDESIGN

Based on this trial work, a number of design alterations were recommended for the mold and oscillation system. The following changes were made:

* mechanical oscillator for improved reliability and sinusoidal motion to allow higher oscillation rates, shorter stroke lengths and shorter negative-strip times.

* improved mold water system for higher water velocity (up to 14 m/s) and elimination of contaminants.

* thicker mold tubes and improved tolerances for better alignment and constraint.

* slightly rectangular tube section to obtain square billets (off-setting the squeezing action of the pinch rolls).

* modified mold lubrication plate for improved oil distribution.

* single-taper tubes until optimum tube taper can be established.

The sinusoidal mechanical oscillator was installed during a July, 1987, shutdown and has operated since that time. Molds of the new design were first used in January, 1988, to cast 110 mm billets. Molds of the new design for other square sizes continue to be installed up to the present time. The preliminary results with this system are presented below.

RESULTS OF REDESIGN

Operational Results

The use of the mechanical oscillators significantly increased caster reliability. Mechanical and electrical failures on the oscillators decreased from about one a month to one a year.

The use of a newly designed mold oil lubricant system has improved oil distribution face to face and reduced oil usage. The new system consists of an oil gallery, machined gap and an O-ring seal, rather than a compressible gasket, to give a more uniform oil flow to each face. Tests measuring the lubricant distribution with the new system showed an average deviation of only 12 percent, face to face, from the mean flow rate compared to 38 percent with the old system at the desired flow rate of 30 mL/min (Fig. 10), allowing a reduction in oil usage of about 50 percent without adversely affecting the amount of oil reaching each face.

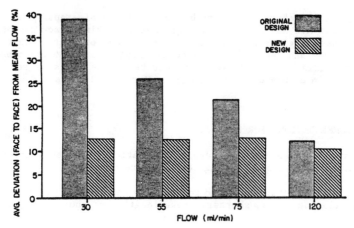

Fig. 10. Mold lubrication distribution - original
          versus new design

Originally, an oscillation rate of 200 cycles a minute and a stroke length of 6 mm were adopted to obtain a negative-strip time of 0.06 s (Fig. 11). The results were rapid mold table bushing wear and oscillator support-beam vibration. After about three months, some mold table movement and billet sticking could be observed. The wear and beam vibration effectively shortened the stroke length and gave an operation with no negative-strip time. Calculations of negative-strip time for various stroke lengths and oscillation rates illustrates this effect (Fig. 11). A reduction in effective stroke length from 6 mm to 5 mm, at 200 cycles per minute, reduces negative-strip time to zero.

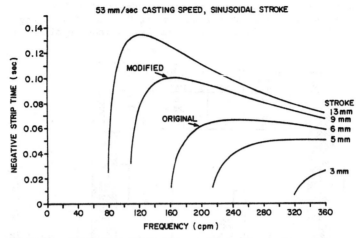

Fig. 11. Negative strip time versus frequency

To reduce the sensitivity to bushing wear, the stroke length was increased to 9 mm and the oscillation rate was reduced to 160 cycles a minute. A change in effective stroke length from 9 mm to 6 mm due to wear or vibration is required at 160 cycles a minute to reduce negative-strip time to zero, an increase of three times the allowable tolerance. A short negative-strip time (0.10 s) is still obtained compared to that with the hydraulic oscillator (0.23 s minimum). The mechanical oscillator was changed to this mode of operation in March, 1988. In addition, the oscillator support beam was stiffened and a program to rebuild mold tables and monitor bushing wear was initiated. These changes reduced the incidence of sticking. Work is continuing to find the optimum stroke length, oscillation rate and negative strip time.

Quality

Operation of the new oscillators at both 6 mm and 9 mm stroke length has resulted in reduced oscillation mark depth. This is most noticeable on the low-carbon grades where the average oscillation mark depth has been reduced by roughly 50 percent (from about 0.50 mm to 0.25 mm).

The shallower oscillation mark depth has resulted in a reduced incidence of transverse billet cracks. Data taken from 15 cm long, low-carbon billet samples collected on each heat for shape and quality control over periods of both 6 mm and 9 mm stroke length operation (0.06 and 0.10 s negative-strip times) show 97 percent free from transverse cracks, 0 percent with cracks of depth greater than 1 mm and 3 percent with cracks shallower than 1 mm (Fig. 12). With the previous hydraulic oscillator and negative-strip times of 0.27 and 0.23 s, respectively, 56 and 81 percent of the samples were free from transverse cracks while 15 and 6 percent had cracks deeper than 1 mm and 29 and 13 percent had cracks shallower than 1 mm. Although the reduction in oscillation mark depth has not been so noticeable on the high-carbon billets, a similar improvement in transverse crack frequency has been observed. Data from a short trial at 0.06 s negative-strip time showed 88 percent of the samples free from transverse cracks, 2 percent with transverse cracks of depths greater than 1 mm and 10 percent with cracks shallower than 1 mm. These results can be compared with those obtained before the new oscillators were installed where 43 and 61 percent of the samples were free from transverse cracks, 16 and 0 percent had cracks deeper than 1 mm in depth and 40 and 39 percent had cracks shallower than 1 mm for the 0.27 and 0.23 s negative-strip time periods, respectively. This is a significant improvement in billet quality.

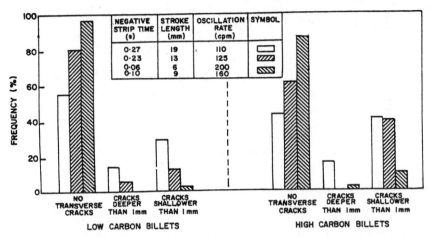

Fig. 12.   Transverse crack frequency - new oscillator

Data collected to date on the 110 mm billet size show that the new mold design has reduced billet rectangularity. On this size, tube width was increased by 2.5 mm on the sides adjacent to the pinch rolls and has reduced average rectangularity from 2.2 mm to 1.2 mm, essentially eliminating billets scrapped for rectangularity beyond the control level of 4.2 mm.

Although billet rhomboidity may be less of a problem on a straight mold caster, the effect of the new oscillator and the thicker, single-taper tubes in the new design is being monitored. On the 110 mm billet size (13 mm thick tubes, 0.9 percent/m taper), where the most data have been collected, no significant difference in mean rhomboidity has been seen among the periods with the new oscillator and the thinner tubes, the new oscillator and the thicker tubes and the earlier period with the old oscillator and the thin tubes. Mean rhomboidity has remained below 1.7 mm in all three periods.

However, an unusual phenomenon has been noted after the introduction of the thicker tubes. Rhomboidity has decreased as tube life has increased. With tube life less than 10 heats, rhomboidity has averaged 2.3 mm, with tube life between 10 and 50 heats, rhomboidity has averaged 1.9 mm and with tube life greater than 50 heats, rhomboidity has averaged 1.0 mm. A corresponding change in tube dimensions has been noted. An increased tube taper has been measured in the region below the meniscus and a reduced taper has been measured in the region at the center of the tube. This situation is considered to relate to improper matching between the shell shrinkage and tube taper. Further work is planned to help understand and correct this problem.

Breakouts

Breakouts are affected by many interrelated factors. Previous studies at Edmonton Works have noted the effect of steel chemistry variables such as carbon content, manganese/silicon ratio, manganese/sulfur ratio and phosphorus level and operational variables such as steel temperature control, tundish design, starter bar design, and others on breakout frequency. Changes in these factors make it difficult to detect changes in breakout frequency due solely to the mold and oscillator improvements.

On the largest billet size (200 mm) changes were made to the starter bar head setup very soon after installation of the new oscillator. An improved starter bar anchoring system was used to reduce the occurrence of breakouts at the start-of-cast due to bolt pullouts. The combination of the new oscillating system and the new starter bar head design has reduced average breakout frequency from 6.7 percent in the 12 months prior to these changes to 2.0 percent in the four months after these changes. In the one month that the new mold design has been in service, breakout frequency has dropped to 1.0 percent.

On the next largest billet sizes (120 mm and 150 mm), only the new oscillator has been used; the variability of the breakout frequency, month to month, has dropped but the average breakout frequency has remained at about the same level. The use of the new mold design on these sizes is expected to result in lower breakout frequency.

On the smallest billet size (110 mm) where both the new oscillator and the new mold design have been in use for the past three months, no significant reduction in breakout frequency has been seen. The changes in mold tube dimensions and billet rhomboidity with increasing mold life on this size suggests incorrect mold taper, which can adversely affect breakout frequency. Also, the billet sticking problems observed with the 6 mm stroke length, 200 cycles a minute oscillation rate operation indicates the sensitivity of this size to oscillation characteristics. Further work is required to understand this problem. Plant trials are currently being carried out with alternative mold tapers and oscillation characteristics to select optimum values for reduced breakout frequency.

SUMMARY

An improved mold oscillation system has been implemented on the Edmonton Works caster after plant trials which included tests with an instrumented tube. Changes were made to the mold design to provide more rigid support of the copper tubes, to allow the use of thicker tubes and to provide higher water velocities with more uniform distribution. A mold lubricating system has been developed which gives improved oil distribution on each face and a mechanical, sinusoidal, oscillation system has been installed which accommodates short stroke length, short negative-strip time, high oscillation rate operation. Preliminary results from this system have shown improved operation owing to reduced mechanical and electrical delays. The new oil lubricant system has allowed a 50 percent reduction in oil consumption. Short stroke, short negative-strip time, high oscillation rate operation has resulted in reduced oscillation mark depth and improved billet surface quality. Further work is being carried out to implement the optimum mold taper and oscillation characteristics for reduced breakout frequency.

REFERENCES

1.   Bakshi, I. A., E. Osinski, I. V. Samarasekera, and J.K. Brimacombe (1988). Measurement of oscillation marks on continuously cast billets with an automated profilometer. <u>CIM Symposium on Direct Rolling and Direct Charging</u>.

2.   Samarasekera, I. V., and J. K. Brimacombe (1982a). The influence of mold behavior on the production of continuously cast steel billets. <u>Met Trans B, 13B,</u> 105-116.

3.   Samarasekera, I. V., and J. K. Brimacombe (1982b). The thermal and mechanical behavior of continuous casting billet molds. <u>Ironmaking and Steelmaking, 9,</u> 1-15.

DESIGN CONSIDERATIONS IN METALLURGICAL PERFORMANCE
OF BILLET CASTER TUNDISHES

D. J. Harris[1], L. J. Heaslip[2], K. E. O'Leary[1], R. W. Pugh[1], C. Jager[3]

[1]Steltech, 1375 Kerns Road, Burlington, Ontario L7P 3H8
[2]University of Toronto
[3]Stelco Steel, PO Box 2348, Edmonton, Alberta T5J 2R3

ABSTRACT

Tundish design and tundish flow modification in multi-strand billet casting systems are discussed with regard to the factors which influence cast metal quality. The combination of physical models of fluid flow (water modeling), which predict flow patterns and fluid residence, and mathematical models, which describe air entrainment, inclusion generation and separation, grade changes (mixed chemistry) during sequence casting, and casting temperature of each strand, is shown to allow the creation of general rules for tundish design in terms of volume, depth, and overall shape. In addition, the optimization of the thermal (temperature and strand uniformity), physical (mixing and residence time), and chemical (oxide contamination from the tundish environment) factors by the placement of flow modifiers (dams, weirs, and baffles) is discussed with reference to maximizing billet quality. Specific examples are drawn from in-plant quality improvement programs.

KEYWORDS

Billet quality, tundish design, flow modification, mathematical model, water model.

INTRODUCTION

The tundish acts principally as a buffer between ladle and mold during continuous casting, distributing flow among multiple strands, at a constant ferrostatic head, and providing a reservoir during ladle exchanges when sequence casting. More recently its important role in determining cast metal quality has been recognized (McLean, 1988) and the technologies of tundish metallurgy are being developed. These technologies, which include flow-pattern control by means of strategically located dams and weirs, alloy and gas injection, liquid metal heating, and filtration have as a prerequisite suitable tundish design in terms of overall shape, ladle stream placement versus exit nozzle locations, and metal depth.

Increasing metal depth in the tundish provides a simple and yet invariably effective means of improving the flotation (i.e. removal) of inclusions through increased fluid residence time. With adequate normal operating depth, casting speeds can be maintained during ladle exchange without a consequent reduction in depth below the critical value

19

(which depends on casting speed and tundish design in the region of the outlet nozzle) at which a draining vortex, with subsequent contamination of the steel, can occur.

In the case of multistrand tundishes (three to six strands being common for billet or bloom casting), however, increased depth leads to increased difficulty in maintaining consistent performance of each strand. Most obvious are the differences in pouring temperatures observed from strand to strand. In addition, stream-to-stream variations in surface roughness or stream quality, inclusion level, gas content, and nozzle-plugging tendency can be observed.

The development of a suitable tundish design not only reflects a wide range of metallurgical considerations but is generally severely constrained by existing plant layout or machine design factors, and operational considerations related to refractory installation and life, deskulling etc.

In multistrand billet/bloom casters with typical tundish design such as those shown in Fig. 1, the nearest tundish outlet nozzle will be less than 0.6 m from the ladle stream entry shroud while the farthest nozzle might be more than 3 m away.

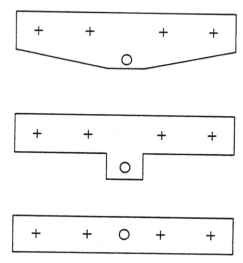

Fig. 1. Some typical tundish shapes
         for billet/bloom casters

The steel arrives at the nearest, or inside, nozzles via a very short path from the inlet point and, therefore, is provided with less time to lose heat and to float out inclusions. In addition, the turbulence created by the momentum of the ladle stream extends over this short path, disrupting the exit stream and carrying down contaminated steel from the tundish surface. The inside strands, therefore, run hotter and dirtier than the outside strands and have poorer stream control. The usual outcome is higher breakout frequency on the inside strands and increased inclusion levels and porosity in the steel that is cast, while the outside strands experience greater difficulty with skulling and maintaining good casting streams.

PHYSICAL MODELING: PRINCIPLES

The physical models were constructed entirely of clear acrylic with internal geometries precisely modeled to the internal geometries of the tundishes being investigated, with geometric scale factors being in the range of one-half to full scale.   They were supplied with a steady water flow through a submerged ladle shroud of equivalent scale. The outlet and inlet flow rates were chosen to satisfy a Froude Number criterion for dynamic similarity, in which case gravitational and inertial forces are considered to be predominant (Heaslip, 1983).

Experiments were conducted with a dye tracer (potassium permanganate) injection for flow visualization and residence time determination with buoyant particle injection (hollow glass microspheres) for inclusion simulation.

CASE STUDIES

Two cases are examined.   In the first case, the overall tundish design is not altered and the capability of flow control inserts to optimize fluid flow performance within that design (symmetric Tee) is examined.   In the second case, an asymmetric Tee-shape tundish is altered to a more suitable design called the Delta-Tee and the performance of this design is examined.

McMaster Works - Tee Design

A schematic diagram of this tundish is shown in Fig. 2.

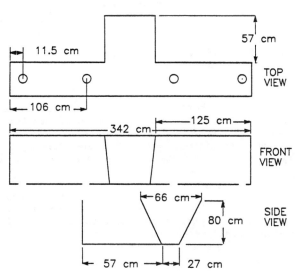

Fig. 2.   Schematic full-scale 12-ton tundish
          water model: Tee design

For all trials, the model was operated in the flow-through mode, simulating a casting speed of 3.8 m/min through a 14 mm ID metering nozzle. The liquid depth in the tundish was maintained at a constant level of 710 mm for all trials. The ladle-to-tundish stream was shrouded and submerged 100 mm below the liquid surface.

After equilibrium had been sustained in the system for 15 minutes, a dye tracer, potassium permanganate ($KMnO_4$), was injected into the system and the progression of fluid flow was recorded through a series of photographs. Minimum retention time was determined by timing from the initial dye injection until the dye first appeared in the mold.

Several dam and weir configurations were used to control flow, and retention time data were generated for these combinations and compared to that with no flow control.

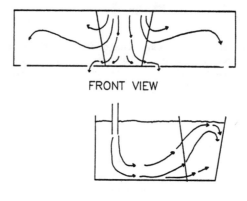

FRONT VIEW

SIDE VIEW

Fig. 3. Flow pattern - no flow
         controls

With no flow controls, the fluid is short circuited from the ladle nozzle to the two inner strands, but it is much slower in reaching the outer nozzles. The minimum retention time for the inner strands was 10 seconds compared to 23 seconds for the outer strands. Also, surface conditions were turbulent, making it difficult to maintain an even slag layer, which in the steel system could result in excessive heat loss and reoxidation.

As indicated in Table 1 and Fig. 4, using configuration 6, with a weir in the pour box and slotted dams located inside the inner nozzles, yielded the highest minimum retention times, the lowest retention time difference between inner and outer strands, and the greatest increase in overall retention.

TABLE 1 MINIMUM RETENTION TIME WITH VARIOUS DAM, WEIR COMBINATIONS

Configuration	Distance from Bottom and Pouring Streams (mm)*					Minimum Retention Time (sec)		% Diff Outer Inner	% Increase	
	a	b	c	d	e	Inner	Outer		Inner	Outer
1	No Flow Control					9.6	23.0	140	0	0
2	-	-	250	450	-	13.2	28.1	113	38	22
3	300	350	-	-	-	18.2	-	-	90	-
4	500	350	250	350	-	31.7	59.6	88	230	159
5	250	350	500	350	-	17.6	30.3	61	83	32
6	-	-	300	400	450	71.0	81.8	15	640	256
7	250	300	-	-	350	28.1	59.7	112	193	159
8	300	320	-	-	400	17.7	21.6			
9	-	-	-	-	350	19.5	79.3			
10	500	300	250	250	350	45.8	90.3			
11	250	350	450	280	420	37.1	54.9			
12	250	350	500	350	350	41.4	78.5			

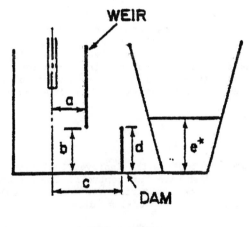

SIDE VIEW

Fig. 4. Flow control placement
(refer Table 1)

A schematic of the fluid flow with this dam and weir configuration is shown in Fig. 5.

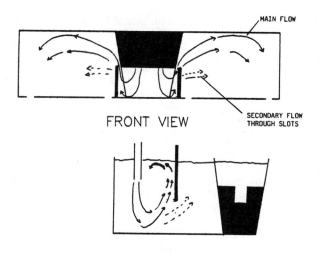

FRONT VIEW

SIDE VIEW

Fig. 5. Flow pattern - optimum
         configuration

The weir in the pour box recirculates the fluid and reduces its velocity before it enters the trough section of the tundish. It also contains the turbulence developed by the ladle stream within the pour box area leaving the slag layer in the tundish trough undisturbed. The dams in front of the inner nozzles force the fluid to the surface, maximizing contact between the molten steel and the slag cover and minimizing the effect of fluid streamlines directly exiting the inner nozzles (i.e. temperature and inclusions)

Plant Trials

Plant trials were initiated to confirm the improved inclusion removal predicted for the 10.9 tonnes tee-shaped tundish with flow controls. The dam and weir flow modifiers were fabricated from tundish board material (65 percent MgO) and set up in the configuration established in the water model studies.

A preliminary trial, carried out on two groups of sequence-cast heats of medium carbon steels (Grades 1141, 1541 and 400) established test procedures and indicated enhanced and consistent inclusion removal in the flow control tundish group (Fig. 6).

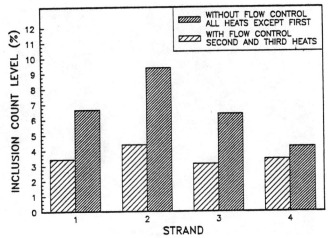

Fig. 6. Inclusion analysis McMaster Works
        billets

Inclusion content with the flow control tundish sequence was reduced to about 50
percent of the level observed on the no flow control tundish sequence (Fig. 6). Data
from the first heats on each sequence was excluded because of delays in installing the
ladle-to-tundish shrouds. Data from the last heat on the flow control tundish sequence
was excluded because of weir erosion.

Visual inspection of the tundish during the casting operation confirmed the improved
flow patterns seen in the water model work. Exposed steel on the flow control tundish
was restricted to the pouring box area (Fig. 7 and 8). Also, temperature measurements
taken along the tundish trough confirmed the uniform metal distribution. A temperature
variation of only 3°C was observed in the flow control tundish, compared to 14°C in
the non-flow control tundish.

Fig. 7.   View of tundish barrel and pour box
during tundish trial

Fig. 8.   View on tundish pour box
during tundish trial

A second trial was conducted to confirm the results of the preliminary trial on two different groups of sequence-cast heats of the same grade. To minimize the number of samples, only strands 1 and 2 were used in the second comparison. The results of the second trial (Table 2) confirm the improved steel cleanness reported from the first trial.

TABLE 2   A COMPARISON OF AVERAGE STEEL
CLEANNESS DURING THE SECOND TRIAL
FOR STRANDS 1 AND 2

Strand	With Flow Control Inclusions/mm^2	No Flow Control Inclusions/mm^2
1	0.008	0.012
2	0.008	0.016

Modifications to the Flow Control Configuration

The performance of the flow control devices is strongly influenced by their service life and the casting parameters. The service life of the dams and weir in the preliminary trial was poor; the bottom of the weir was flush with the steel level after the third heat in the sequence-casting of five heats. Although the average number of heats in sequence is 2.8 in 1988, the service life of the dams and weir must be greater than five heats, on average, to provide effective inclusion removal for more than 70 percent of the heats produced.

To achieve the high service life of the dams and weirs, their shapes and materials have been modified (Fig. 9) for an upcoming trial. The weir will be fabricated in two sections; the upper section will be made of 65 percent MgO at 93 lb/ft^3 and the lower section will be made of 85 percent MgO (high strength) at 155 lb/ft^3. The lower section will be subjected to high fluid flow velocity (Fig. 3).

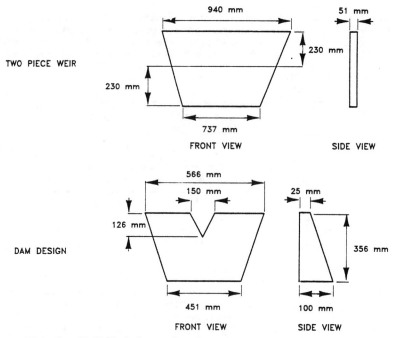

TWO PIECE WEIR

940 mm

51 mm

230 mm

230 mm

737 mm

FRONT VIEW   SIDE VIEW

DAM DESIGN

566 mm

150 mm

126 mm

25 mm

356 mm

451 mm   100 mm

FRONT VIEW   SIDE VIEW

Fig. 9. Modified dam and weir

Tundish Stream Temperature Deviation

Measurement of residence time distribution for each outlet nozzle in the model allows calculation of the true mean tundish residence time for a particular strand i ($\bar{t}_i$) by means of an integration of the data according to Equation 1:

$$\bar{t}_i = \frac{\int_0^I C\tau \, d\tau}{\int_0^I C \, d\tau} \qquad\qquad \text{Equation (1)}$$

where: I = upper integration time limit
        C = normalized tracer concentration
        τ = time

Conversion of these residence times to temperature loss for a particular strand requires accounting for the thermal losses in the tundish. The pouring temperature (T_i) of a strand (i) at any time, (τ), is given by:

$$T_i - T_L [\tau] - \bar{t}_i [\dot{Q}_{s,r}(\tau)] \;/\; \rho.V.C_p \qquad\qquad \text{Equation (2)}$$

where: T_L - ladle stream temperature

$\dot{Q}_{s,r}$ - rate of heat loss from tundish to surface and refractory

$\rho.V.C_p$ - total tundish specific heat

and the deviation from stream-to-stream $(T_i - T_j)$ is given by:

$$T_i - T_j - (\bar{t}_i - \bar{t}_j) \dot{Q}_{s,r} [\tau] \;/\; \rho.V.C_p \qquad\qquad \text{Equation (3)}$$

Thus, stream-to-stream temperature deviation is exacerbated by large differences in true mean residence times for each strand and when the rate of heat loss in the tundish is large, such as during tundish filling.

TABLE 3   PREDICTED TEMPERATURE DIFFERENCE
- MCMASTER WORKS TUNDISH

	Δt inside - outside (°C)
No flow controls	+21.4
Optimum flow controls	+4.8

These predictions compare favorably with the in-plant measurements reported previously in this paper.

Edmonton Works - Delta-Tee Design

The original tundish design at EW is shown in Fig. 10. This arrangement is referred to as an asymmetric-Tee design, and is subject to extreme variations in strand performance (Table 4) since each strand behaves in a unique manner. Therefore, it was decided to consider a new design which allows for improved symmetry by moving the ladle pouring position to one which is equi-distant from each of the outside strands (Fig. 11).

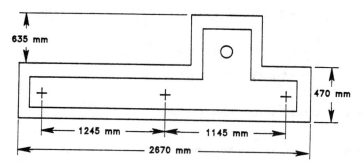

Fig. 10. Original assymetric Tee design - EW

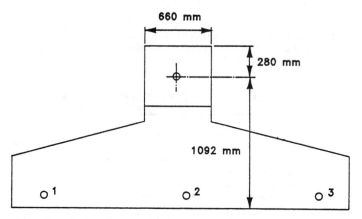

Fig. 11. Delta-Tee tundish - EW design

Fluid Flow Pattern - No Flow Control

In the absence of flow controls, the fluid flow pattern showed very short residence time (short circuit flow) for the fluid exiting through the inner strand (No. 2). The results are presented in Table 4.

TABLE 4 MINIMUM FLUID RETENTION TIME - NO FLOW CONTROLS

Tundish Design	Tundish Depth (mm)	Minimum Retention Time (sec)		
		Strand No. 1	Strand No. 2	Strand No. 3
Assymetric-Tee	356	18	8	45
Delta-Tee	508	34	14	39

The fluid flow retention time for Strand No. 2 is 61 percent lower than the averages for Strands No. 1 and No. 3 for the Delta-T tundish without flow controls.

Flow Controls - Delta-Tee Tundish

The effect of flow control, i.e. dam and weir, on fluid retention time was evaluated. Several configurations of flow obstacles were investigated. A schematic of the Delta-Tee tundish with flow controls is presented in Fig. 12 and Fig. 13 and preliminary results are summarized in Table 5.

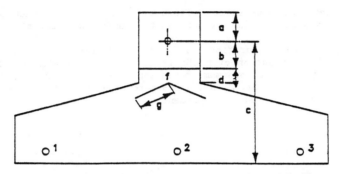

Fig. 12.   Delta-Tee tundish configuration with flow
            controls

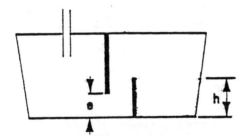

Fig. 13.   Side view, Delta-Tee
            tundish

TABLE 5 MINIMUM FLUID RETENTION TIME - FLOW CONTROLS
          - DELTA-TEE TUNDISH

Configuration	Tundish Depth (mm)	Weir Position (mm)					Dam Position (mm)				Minimum Retention Time (sec)		
		a	b	c	d	e	f	g	h	Angle (°C)	Strand No. 1	Strand No. 2	Strand No. 3
1	508	280	250	1092	125	125	152	500	200	105	41	43	41
2	508	280	250	1092	125	125	152	500	152	105	39	41	43
3	508	280	250	1092	125	125	152	500	200	135	44	46	47

The dam and weir increased the minimum fluid retention time of each strand. The results indicate very similar behavior between the inner and outer strands.

The optimal flow controls configuration (Configuration 3, Table 5) shows a significant improvement in fluid distribution over the Delta-Tee tundish without flow controls (Fig. 14). The increase in minimum retention time of each strand for this configuration is 23 percent on Strand No. 1, 69 percent on Strand No. 2 and 17 percent on Strand No. 3 (overall increase 36.4 percent).

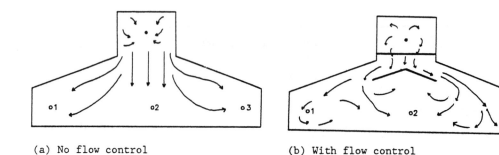

(a) No flow control   (b) With flow control

Fig. 14. New Delta-Tee shape - EW

This arrangement indicates an average residence time of 45.6 sec, compared to 29 sec, for no flow controls, with Strand No. 2 increasing the most dramatically.

To quantify the effect of this flow control (dam and weir) configuration on inclusion separation in the Delta-Tee tundish, experiments were carried out on the one-half scale water model of the Edmonton Works tundish.

Equipment and Design

Fig. 15 is a schematic drawing of the experimental arrangement used for the inclusion separation study.

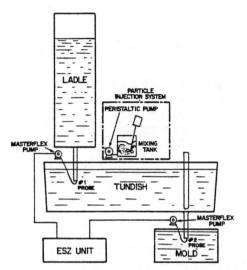

Fig. 15. Schematic of the one-half
          scale Delta-Tee tundish
          experimental arrangement

For simulating inclusions in molten steel (Wilshynsky, 1986), hollow glass microspheres with an appropriate number density of 2.5×10^8 particles/m^3 in the size range of 50 to 120 microns were fed continuously, at a constant rate, into the tundish through the inlet shroud. The specific density of these glass microspheres was 340 kg/m^3.

A monitoring system (LIMCA technique), based on the Coulter Counter technique, in which nonconductive particles pass through an electrically insulated orifice in the presence of an electric current, was used for counting and sizing the "inclusions".

Particle Measurement Technique

The principle of the LIMCA (Nakajima, 1986) technique (resistive pulse technique) is shown in Fig. 16.

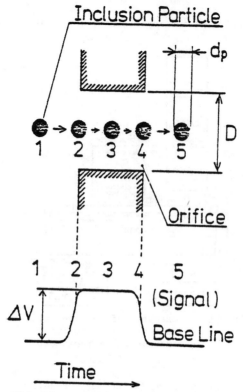

Fig. 16. Principle of particle
          detection

When entrained particles pass through an orifice through which electric current also
passes, associated changes in electrical resistance cause voltage pulses. These
voltage signals are detected using an oscilloscope.

The height of each pulse is then measured and stored in the Multi-channel Analyzer.

Fig. 17 shows the conversion of this information to particle number density versus
particle size.

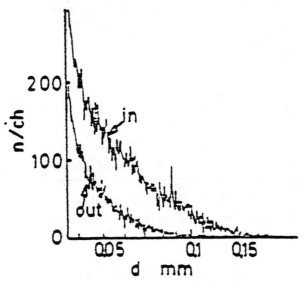

Fig. 17.   Size distribution of inlet and
           outlet particles - model of
           tundish

The upper curve provides the number and size distribution of particles being fed into
the tundish, while the lower curve represents the number of such particles exiting
into the mold.

Inclusion Separation - No Flow Controls - EW

Several tests were conducted to determine the inclusion separation capability of the
Delta-Tee tundish with no flow control devices. The inclusions carried into the mold
(without flow controls) are presented as percentages in Fig. 18.

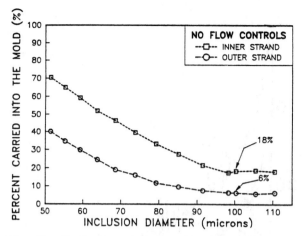

Fig. 18. Inclusion size distribution - no flow
          controls ("steel" depth 508 mm)

With no flow controls, the size distribution entering each mold is significantly
different.   Eighteen percent of 100 microns diameter particles are carried into the
inner strand compared to six percent into the outer strand.

Increasing the tundish operating level (Fig. 19), from 508 mm to 710 mm, showed an
improvement in flotation but little improvement in homogeneity between strands.

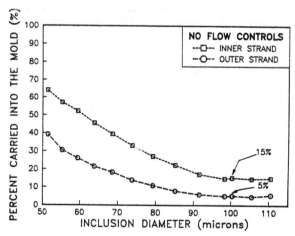

Fig. 19. Inclusion size distribution - no flow
          controls ("steel" depth 710 mm)

Inclusion Separation - Flow Controls

The effect of flow controls, dam and weir, on inclusion separation was also evaluated.
Fig. 20 presents the results.

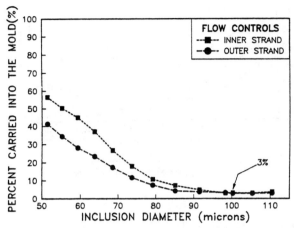

Fig. 20.   Inclusion size distribution - flow
           controls ("steel" depth 508 mm)

The flow controls ensure the benefit of strand-to-strand uniformity in inclusion
separation. The flow modification reduced the overall inclusion carryover (100 microns
size) to three percent.

Increasing the tundish operating level to 710 mm, from 508 mm, reduced the carryover
of 50 to 80 microns inclusions by 18 percent in the inner strand and by 11 percent in
the outer strands, Fig. 21.

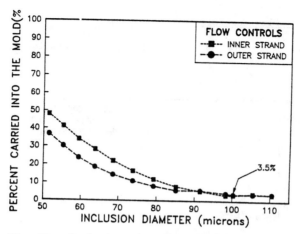

Fig. 21. Inclusion size distribution - flow
          controls ("steel" depth 710 mm)

PLANT TRIALS - EW DELTA-TEE

The Delta-Tee tundish has been used to cast twelve heats to date (four medium and three
high carbon, Si-killed heats; five aluminum-treated). Visual inclusion counting on
hot etched billet samples from the silicon-killed heats indicates a reduction in
inclusion level on Strand 1 and 2 and about the same level on strand 3 compared to
similar grades cast in the regular assymetric Tee-shaped tundish (Fig. 22).

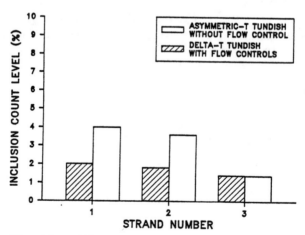

Fig. 22. Billet inclusion count Edmonton Works

Inclusion counting on samples from the aluminum treated heats is still ongoing. Heats of similar chemistry were not cast in the regular assymetric-Tee-shaped tundish, and so no comparison data is available.

Pinholes and blowholes were monitored on the medium carbon heats since historical data exists. A reduction in pinhole level was observed on strands 1 and 2 on the two delta tundish heats compared to the levels measured on eight previous medium carbon heats cast in the regular assymetric-Tee-shaped tundish. Blowhole levels remained low with both tundishes (Fig. 23).

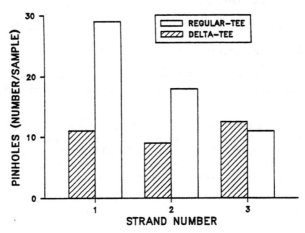

Fig. 23. Billet pinhole count - Edmonton Works

Tundish Steel Temperature - EW

Using the residence time data generated with the water models and the mathematical analysis described earlier, predictions were made concerning the stream-to-stream temperature variation anticipated with the Assymetric-Tee and the optimized Delta-Tee designs (Table 6). Since the Delta-Tee is not yet used in regular production, temperature deviations on the Assymetric-Tee were used to confirm the applicability of the models.

TABLE 6   STRAND TEMPERATURE VARIATION

Tundish Design	Temperature Variation (°C)*	
	Strand 2	Strand 3
Assymetric-Tee (Predicted) (No Flow Control)	-2.0	-22.6
Assymetric (Plant)	-4.0	-16.7
Delta-Tee (Predicted) (Optimum Flow Control)	-2.0	-2.0

* Relative to Strand Number 1 (shorter residence time)

The predictions and plant measurement are in good agreement. A large stream-to-stream temperature deviation is found to be associated with the Assymetric-Tee design. The model indicates that this variation can be nearly eliminated with optimum flow controls in the Delta-Tee design. This improvement implies fewer breakouts and greater homogeneity in billet quality with the Delta-Tee design.

SUMMARY

A symmetrical tundish design that maximizes fluid retention time using optimum flow control is a prerequisite for consistent tundish performance. This configuration can eliminate strand-to-strand variation in both temperature and quality. Combined use of a water and mathematical model is a viable technique to optimize tundish design and flow control.

REFERENCES

McLean, A. (1986), Howe Memorial Lecture, The Turbulent Tundish Contaminator or Refiner?, Vol 7, Steelmaking Conference Proceedings. ISS-AIME Society, Toronto, Canada, pp 3-23.

Heaslip, L. J., A. McLean, and I. D. Sommerville (1983), Continuous Casting, Vol 1, ISS-AIME, pp 70-71.

Wilshynsky, D. O., D. J. Harris, and L. J. Heaslip (1986), "Water Modeling of Nonmetallic Inclusion Separation in a Steel Casting Tundish", Fachberichte Hutte Metal, Vol 23, No. 4.

Nakajima, H., F. Sebo, S. Tamaka, L. S. Dumitru, D. J. Harris, R. I. L. Guthrie (1986), "On the Separation of Nonmetallic Inclusions from Tundishes in Continuous Casting Operations - A Water Model Study", 5th International Iron and Steel Congress, AIME, Vol 6, Washington, pp 705-715.

A COMPARISON BETWEEN THE INTERNAL QUALITY OF
POWDER CAST AND OIL CAST BILLETS IN Nb
TREATED STEELS

Eduardo J. Cordova

Hamilton Specialty Bar Division
Slater Industries Inc.
Hamilton, Ontario, Canada   L8N 3P9

ABSTRACT

Trials, to produce powder cast billets, were started at the Hamilton
Specialty Bar Division of Slater Industries in May, 1985.

The trials served to compare, from a metallurgical point of view,
the internal and external quality of powder cast billets versus oil
cast billets. Parameters studied included pinholes, slag spots,
depth of oscillation marks, internal cracking, rhomboidity and
cleanliness for Nb treated steels. Emphasis was placed on the
internal quality comparison.

This paper will discuss, in detail, the results obtained, plus some
of the theoretical aspects governing the formation of oscillation
marks.

KEYWORDS

Oscillation marks; casting powder; casting oil; near surface
cracks; % off-square; billet; heat flux.

INTRODUCTION

In line with the commitment to continuously improve on quality, the
Hamilton Specialty Bar Division commenced powder cast trials in May,
1985. The benefits originating from submerged powder casting have
been thoroughly documented in the literature. Most of the relevant
information, however, dealt with slab and bloom casters, with little
information on billet casters.

Similarly, during the information gathering stage of the trials, it
was found that most of the readily available powders had been
originally formulated for slab or bloom casters. Consequently,
operational and quality problems, such as, low powder spreadability,
bridging and too high viscosity, became evident when using these
powders.

These powders, therefore, never properly performed their intended functions of insulating the steel, protecting it from reoxidation and inclusion absorption. The smaller mould size in our billet caster also contributed to other problems; i.e., the formation of steel-oxide-powder flux particles in the billets, particularly during start up of the cast.

Improved results were obtained after several trials which included the use of start up powders and re-formulation of powders to better suit billet casting conditions. A casting powder was obtained that fulfilled the requirements of our particular operation.

It is not the scope of this paper to discuss the details of the powder selection and its performance, but to concentrate on the quality comparison between powder cast and oil cast billets, after the best powder casting practice had been established at the HSBD of Slater Industries Inc.

FACILITIES

The Hamilton Specialty Bar Division has an annual melting capacity of approximately 350,000 tons. The melting unit is a 60 ton EAF equipped with a submerged taphole. Subsequently, the steel is refined in our ladle refiner and then cast in one of the two 3-strand billet casters. The main features of the Melt Shop are summarized in Table 1.

TABLE 1   Equipment Main Features

i)   EAF:

Capacity: 60 tons
Submerged tap hole
Transformer rating: 35 MVA
Water-cooled roof and panels
Manufacturer: American Bridge
                    (1967 commissioning)

ii) Ladle Refiner:

Capacity: 60 tons
Arc heating unit with stirring
Transformer rating: 15 MVA
Heating rate: 5.2 deg C/minute
Stirring: Ar gas (bottom directional plug)
Manufacturer: Midland Ross
                    (1985 commissioning)

iii) Billet Casters:

Mould size 127mm X 178 mm (5 in X 7 in)
3.96m radius (13 ft)
3 strands
Manufacturer: Concast (1968 and 1970
                            commissioning)

iv) Grades Produced:

Carbon AISI 1010 to 1090
Resulphurized AISI 1100's 1500's series
Alloy AISI 5100's, 8600's, 4000's series

v)   Rolled Products: Flats, Rounds and special sections

TRIAL SET-UP

The casting conditions for the trials were summarized as shown in TABLE 2. In agreement with the commonly known fact that the depth of oscillation marks decreases with increasing mould oscillation frequency,f, and decreasing stroke length, h, 2.3 Hz (140 cpm) and 6.35mm (0.250 in) were chosen for both powder and oil cast billets.

AISI 5160 grade was chosen because it represented a large part of our melting grades, and trial heats could be set up regularly.

The comparison was based on normally monitored parameters, such as, internal cracking, rhomboidity and microcleanliness. Billets were also scarfed to evaluate subsurface pinholes and slag spots. The surface of the billets was evaluated by measuring the depth, d, and pitch, p, of the oscillation marks.

To eliminate problems by the possible flattening of the oscillation marks by the withdrawal rolls on the wide face, the depth of the oscillation marks was measured on the billets narrow face, using metallographic specimens.

TABLE 2   Trials Casting Conditions

Parameters Practice	Vc m/min (in/min)	h mm (in)	f Hz (cpm)	Mould size mm (in)	C %
Powder	2.4 (62)	6.35 (0.250)	2.3 (140)	127 X 178 (5 X 7)	0.58 0.60
Oil	2.4-2.6 (62-65)	6.35 (0.250)	2.3 (140)	127 X 178 (5 X 7)	0.58 0.60

RESULTS AND DISCUSSION

The most obvious characteristic was the difference in appearance of the oscillation marks (Fig. 1). The oscillation marks in powder cast billets were more evenly spaced and symmetrical (Fig. 2). The depth of the oscillation marks, d, was plotted against the pitch, p. In general, it was found that the depth increased with increasing pitch, and that the oscillation marks were deeper in powder cast billets as shown in Fig. 3.

In Fig. 3, it is worth noting the following: Casting speed has such an influence on the depth of the oscillation marks, that mould level fluctuations can vary the strand to mould relative motion, creating greater variance under so called constant casting conditions (3). Moreover, casting speed variations can induce deeper oscillation marks, with a shorter pitch (2). These factors 'debilitated' the depth-to-pitch relationship found in oil cast billets, since we exercised better casting speed and mould level control in powder cast billets (Fig. 4).

To better explain the differences in oscillation mark appearances, between oil and powder cast billets, the mechanism of oscillation mark formation must be understood.

Several mechanisms have been proposed in the past. A "tearing and healing" mechanism was proposed by Savage and Pritchard and, later, by Sato (1). Here, the shell sticks to the mould wall, it then 'tears' during the upward motion of the mould, allowing steel to overflow, with a subsequent 'healing' period during the downward motion of the mould. The rather straight appearance of oscillation marks cannot, however, be explained very satisfactorily by such 'rupturing' mechanism.

Emi et al (1), suggested that oscillation marks are formed due to a
"folding of the meniscus shell", caused by the mould movement during
the negative strip period. In this mechanism, the meniscus shell is
folded towards the steel by a liquid flux layer, which in turn is
being pushed downward by a solidified slag layer at the mould wall.
The same study also classified oscillation marks, based on the
presence or absence of subsurface hooks.

More recently, Takeuchi and Brimacombe (1) have suggested that the
formation of oscillation marks and the presence or absence of
associated subsurface hooks, depends on the "generation of pressure
in the flux channel" and whether or not the meniscus skin behaves in
a rigid or semi-rigid fashion. In this mechanism, the rigid and
semi-rigid skins are subject to the same flux pressures in the
channel between the mould wall and the steel. Accordingly, during
the negative strip period, the meniscus skin is pushed toward the
steel by the flux pressure. Once the mould and strand start moving
at the same speed again (i.e., start of positive strip time) the
skin will be pushed back toward the wall by the ferrostatic pressure
and negative flux pressures developing. In the event that the skin
has enough mechanical strength to withstand being pushed back (rigid
skin), the steel will overflow at the weakest sections, giving rise
to a subsurface hook. On the other hand, if the skin behaves
essentially as a liquid (i.e., semi-rigid), it will deform back
toward the wall completely, without any steel overflow and,
therefore, no subsurface hook. The rigidity or semi-rigidity of the
skin will depend on many factors, such as, low superheat, carbon
content and heat flux.

A casting powder will, amongst other things, reduce the heat flux
between mould wall and strand by thermally insulating the strand.
In oil billet casting the heat flux is much higher giving rise to
a more rigid meniscus skin that resists being deformed during the
negative strip time. This results in a shallower oscillation mark.
The lower heat flux in powder casting produces a weaker meniscus
skin which deforms easily, originating a deeper oscillation mark.
Obviously, the relatively lower oil viscosity and the accompanying
lower 'oil pressure in the channel' will also aid in producing
shallower oscillation marks. More research work is required to
fully explain the actual lubrication mechanisms for both powder and
oil casting, so that such differences can be fully understood.

Powder cast billets showed a significant improvement in % off-square
values, when compared to oil cast billets as shown in Table 3. This
could be explained by a more uniform heat extraction in the mould
which resulted in a more even shell thickness as revealed by the
sulphur prints. The more uniform shell thickness reduced the
tendency of the billet to deform in the lower part of the mould
(shell-mould non-contact zone) and after exiting the mould.

TABLE 3 % Off-Square Comparison Between
Oil & Powder Cast Billets

Category	Average	Standard Deviation	Sample Size
Powder	1.30	1.11	37
Oil	3.35	1.65	50

The significant reduction of near surface cracks in powder cast billets can be attributed to the better thermal insulation and more uniform heat flux in the mould, due to the powder slag. Near surface cracks were perpendicular to the billet surface, in their entirety, except when they grew along the line where the dendrites from two sides impinged upon each other. Some cracks exhibited both conditions. In Fig. 5, no trend can be established; however, two populations can be differentiated, with the powder cast billets having the lower cracking average as summarized in Table 4. This can be explained by the way in which near surface cracks form and grow. For instance, it has been reported (5) that cracks are nucleated at very early stages during strand solidification, in a region just below the meniscus. Cracks grow toward the centre of the billet as solidification proceeds, and as the billet exits the mould, cracks already present are made worse. In the case of powder cast billets, the flux film formed decreases the thermal stresses present during shell solidification. In oil cast billets, these stresses are more severe (higher heat flux), giving rise to more cracks at the solidification front. The non-uniform heat flux present between shell and mould in oil cast billets causes further distortion of the billet shape. This distortion encourages the growth of existing cracks.

TABLE 4   Near Surface Cracking Total Length
          (Inches) per Cross Section

Category	Average	Standard Deviation	Sample Size
Powder	0.71	0.62	37
Oil	1.56	0.65	50

Microcleanliness ratings (ASTM E45) revealed a slight overall improvement in powder cast billets as shown in TABLE 5. These values were a reflection on the practices and procedures used during melting, refining and casting. The ability of the powder to obsorb inclusions and retain its fluidity was important for proper lubrication. The ability of the powder to protect the strand from reoxidation was reflected in the lower oxide rating.

TABLE 5   Cleanliness Ratings for AISI 5160
          Si-killed: Powder vs Oil

Practice / Inclusion Type	Type A T	H	Type B T	H	Type C T	H	Type D T	H	Plate II Ox	Si
Powder	1.8	0	0.4	0	2.0	0	0.2	0	0.8	3.8
Oil	1.3	0	1.0	0.2	1.1	0.1	1.1	0.2	2.3	3.3

Subsurface quality was determined by evaluating the frequency and severity of subsurface slag and pinholes. Scarfed billets were rated, using an internally developed rating system, which relates defect frequency and severity to give an overall quality rating. The higher the value, the worse the quality. A value of 40 or less represented acceptable quality.

Initial results indicated a marked worsening of the quality of the powder cast billets. It became evident that the major problem was at the start of the cast, where billet quality was completely unacceptable. After the first ten billets, quality started to improve, but it was still worse than the average quality rating for oil cast billets (Fig. 6).

To improve the situation, a 'start-up' powder was used. Start-up powders melt quicker and provide faster the necessary lubrication at the start-up of cast. This change reduced the 'transition billets' from the first ten to the first five. However, more improvement was necessary. Further improvements on mould level control and casting speed (Fig. 4) were necessary to achieve acceptable quality levels after the third billet. TABLE 6 and Fig. 7 showed the higher quality levels obtained. Our automatic mould level control provided the steadiness necessary to minimize mould level fluctuations. Large fluctuations result in inclusion and flux entrapment during oscillation mark formation, when the steel overflows the meniscus. Further work is necessary to reduce the number of transition billets from three to one.

TABLE 6   Subsurface Rating Comparison Between
            Powder and Oil Cast Billets

Category Practice	Pinhole & Blowhole Severity	Pinhole & Blowhole Frequency	Slag Severity	Slag Frequency	Overall Billet Rating Severity	Overall Billet Rating Frequency
Oil Cast Billets	10.1	7.9	15.7	12.7	27.2	21.1
Powder Cast Billets	5.8	5.5	11.3	11.3	17.2	16.8

CONCLUSIONS

From the trials carried out at the Hamilton Specialty Bar Division, we can conclude that:

(1) Billet near surface cracking and % off square are lower in powder cast billets than in oil cast billets.

(2) Oscillation marks are deeper, more evenly spaced and more symmetrical in powder cast billets.

(3) Mould level fluctuations and casting speed must be controlled effectively to minimize sub-surface defects, such as, slag entrapment and pinholes. The use of a start-up powder helped obtain acceptable quality levels earlier in the cast.

ACKNOWLEDGEMENTS

Thanks to B. Joslin and J. Zanin for their total commitment to the trials, and to B. Bowman for the opportunity to prepare this paper.

REFERENCES

(1) Takeuchi, E., Brimacombe, J.K., "The formation of oscillation marks in the continuous casting of steel slabs", Metallurgical Transactions B. Sept. 1984, vol 15B, pp 439-509.

(2) Wolf, M.M., "On the interaction between mold oscillation and mold lubrication" Mold Powders for continuous casting and bottom pour teeming. I.S.S. 1987, pp 33-44.

(3) Cramb, A.W. Mannion, F.J., "The measurement of meniscus marks at Bethlehem Steels Burn's Harbor slab caster" Mold Powders for continuous casting and bottom pour teeming I.S.S. 1987, pp 83-93

(4) Uchiyama, H., "Characteristics and selection criteria of mold flux in different casting conditions". Paper prepared for AISI Technical Committee on Strand Casting meeting of September 25-26, 1985.

(5) Saucedo, G.I., Gass, R.T., "The formation of longitudinal off-corner subsurface cracks in continuously cast steel billets". Steelmaking conference proceedings, vol 70, Pittsburgh, Pennsylvania, 1987, I.S.S. of AIME, pp 177-186.

(6) Brimacombe, J.K., Samarasekera, I.V., Walker, N., Bakshi, I., Bommaraju, R., Weinberg, F., Hawbolt, E.B., "Mould Behaviour and Solidification in the Continuous Casting of Steel Billets", I.S.S. transactions, vol 5, 1984, pp 71-93.

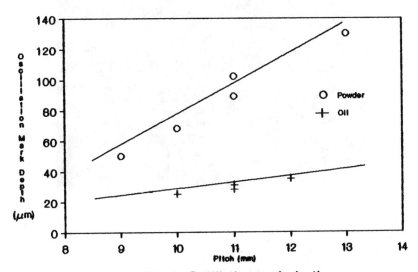

Fig. 3 Oscillation mark depth
versus pitch

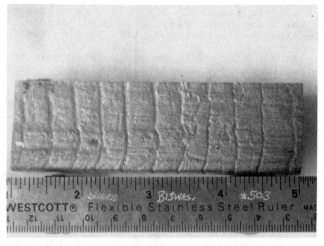

(a)

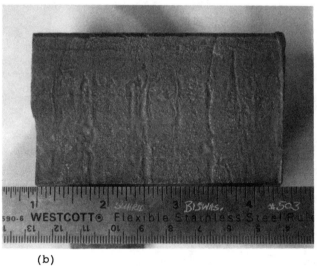

(b)

Fig. 1. Oscillation marks on the surface of
continuously cast billets: (a) Powder
cast and (b) Oil cast.

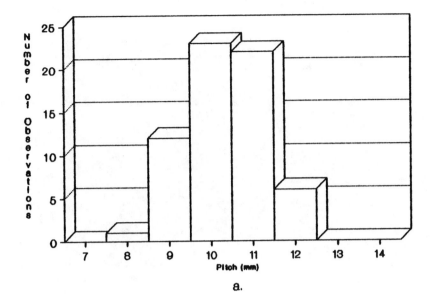

a.

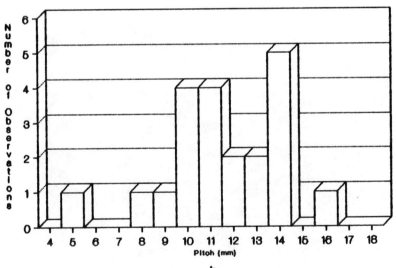

b.

Fig. 2   Pitch distribution in
a. powder cast   b. oil cast

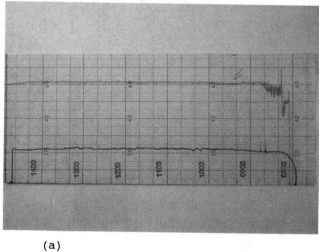

(a)

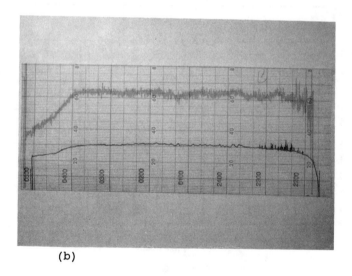

(b)

Fig. 4. Speed charts for (a) Powder cast billets
and (b) Oil cast billets.

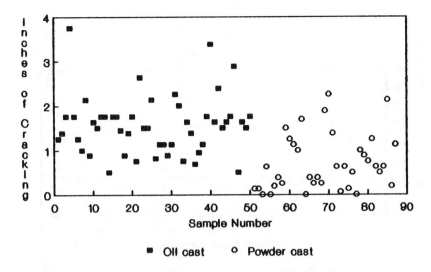

Fig. 5 Internal cracking distribution.

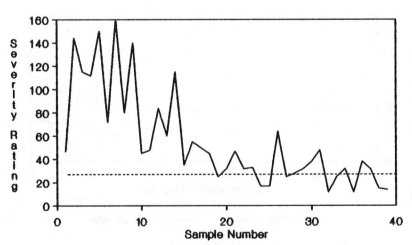

Fig. 6 Billet scarf severity rating
for powder cast (solid line)
avg. for oil cast (dotted line)

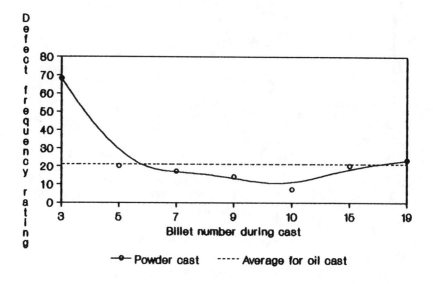

a.

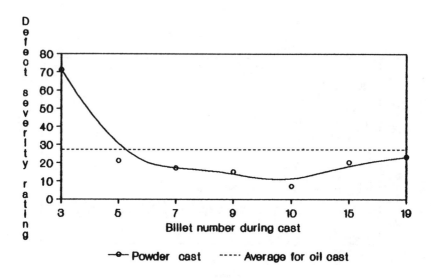

b.
Fig. 7 Scarfed billet defects.
a. Frequency rating b. Severity rating

CONSIDERATIONS FOR THE HOT CHARGING OF BILLETS

John Guerard, Sr. Process Metallurgist
Melting and Casting

Lake Ontario Steel Company
Whitby, Ontario L1N 5T1

ABSTRACT

The process of hot charging is examined from a billet producers' point of view. The advantages of hot charging are reviewed and suggestions are made as to possible methods of quantifying these advantages. Process limitations which could prevent or reduce the benefits of hot charging are also discussed.

Billet quality – the major limitation to achieving significant hot charging benefits – is the main topic of this discussion.

KEYWORDS

Hot charging; cost reduction; billet quality.

INTRODUCTION

The processes of direct rolling and hot charging of billets are both methods to decrease production costs. Hot charging is considered to be a process involving the delivery of billets at higher than ambient temperature to a conventional reheat furnace. Discussion in this paper will be limited to hot charging since it is of most interest to an existing plant such as Lasco, although the considerations should apply to direct rolling as well. The body of the paper is divided into three sections. The first includes benefits of hot charging, the second examines limitations and in the third, implementation methods are discussed. Comments on the importance of hot charging when producing billets are included in the summary.

BENEFITS

Cost reductions can be realized in three areas, energy, production rates and inventories. Although there are interrelationships between the three areas an attempt has been made to assign benefits to one of the areas.

Energy

Energy requirements are reduced by using the sensible heat in the billet as soon as it is cast. The reduction of cost is inversely proportional to the billet charging temperature. (Fig. 1) Decreases in scale loss and decarburization depth also result from higher charging temperatures. Theoretical calculations should be verified by plant trials since furnace efficiencies vary and heat equalization in the billet is also a necessity.

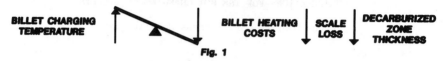

Fig. 1

Production Rate

Production rates are critical to the cost of an operation and hot charging can result in significant increases in rolling rates. The mill rolling rate is generally controlled by the reheat furnace heating rate which is really a function of billet charging temperature. Thus the rolling rate increases with billet charging temperature until some other limiting level is reached. (Fig. 2) It is important to establish these other rate limiting processes on the mill so realistic benefits can be predicted for hot charging.

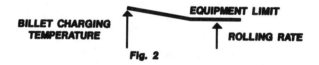

Fig. 2

Inventory

Hot charging makes it possible to decrease the turnaround time on billet inventories. (Fig. 3) This results in lower cost for inventory. A decrease in turnaround time of about two weeks would by typical. There are also cost savings which result from less handling of the inventory. Decreased material handling is an important consideration for a billet producer where there can be hundreds of pieces in one heat. Less handling of the material also decreases the chances of damaging or misplacing the billets.

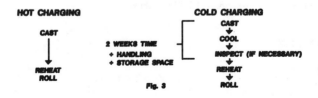

Fig. 3

LIMITATIONS

The benefits possible through hot charging appear to make immediate
implementation seem imperative. However, as with all processes, there are some
limitations which should be considered. These limitations may be particularly
costly to overcome with an existing plant. Discussion of limitations will be
under four headings: Plant layout, timing, temperature and quality.

Plant Layout

The benefits of hot charging may be difficult to realize in an existing plant
where both the direction and distance for material flow can be problems
requiring complex and costly solutions. It may be necessary to implement
insulated transfer cars or enclosed conveyors to maintain high billet
temperature. Although direction and distance may prevent the realization of
100% of the benefits there may still be a cost reduction with a system that is
not ideal.

Timing

Hot charging changes scheduling requirements dramatically. The production
rates of the caster and the mill must now be tied together directly. In an
existing plant this means either the caster or the mill will not be running at
maximum rates. If the caster can produce steel to match the maximum rolling
rate on the mill it will be necessary for the caster to run slower or produce
inventory when the mill is rolling a slower section. This inventory would have
to be reheated. There is a match between the mill and the caster when cold
charging but only as a result of the buffer of billet inventory. If the two
units are tied directly the slowest process will always control the rate. This
is a particularly important point. To aim for 100% hot charging it is
necessary to have a narrow range of product which can be rolled at a consistent
rate or to be able to vary the production rate from the caster to match the
mill. The match between caster and mill must be considered a significant
factor with rolling rate increases of 30% being experienced as a result of hot
charging.

Temperature

The process of hot charging necessitates the handling of materials at high
temperatures. While this does result in energy savings it also creates
significant material handling problems. The majority of billet movement is
done with electromagnets. High temperature (Curie Temperature) results in loss
of magnetization of the billets and slightly lower temperatures can result in
magnet damage. These handling problems can be designed out of a new plant but
may require costly modification to an existing plant such as new conveyors or
transfer cars.

Temperature is also of significance since it must be uniform throughout the
billet before rolling. The temperature distribution in a cast billet is far
from uniform. Thus, devices to maintain and equalize cast billet temperature
must be implemented such as those mentioned in the plant layout discussion.

Quality

The previous three categories are yes/no situations that can be addressed by

engineering, equipment or economic solutions. However, the most critical
consideration is the quality of the billet that will be used to produce the
rolled product and quality does not have a yes/no answer. Rolling a billet of
unacceptable quality quickly cuts into the economic gains made by hot charging
and an assessment of the consequences should be carried out similar to the one
shown in Fig. 4. The control of quality includes identification (or
traceability), chemical analysis, surface integrity and internal integrity.

CONSEQUENCES OF ROLLING A DEFECTIVE BILLET

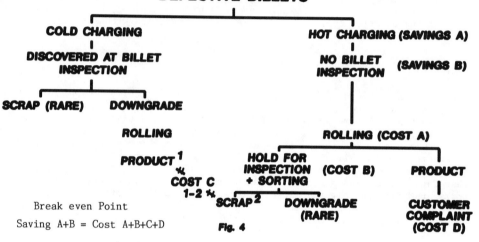

Break even Point

Saving A+B = Cost A+B+C+D

Fig. 4

Mismatches between billet production and rolling rates can be handled by the
periodic introduction of cold billets. This is where the question of billet
identification or traceability becomes important, especially in the case where
hot and cold billets are being alternated. The maintenance of correct billet
identification becomes a stronger quality concern when heats are not rolled in
groups. Chemical composition limitations on grades which can be hot charged
have not been restrictive to most plants. However, timing of the chemistry
results becomes much more critical. The rolling of a heat that is not the
correct chemistry will severely restrict the possibility of finding an
alternate use and will likely result in the product being scrapped. The timing
of sampling and chemistry reporting must be reviewed before hot charging.
Control of chemistry is also critical. If testing indicates that the chemistry
is not suitable for the product it is possible to remove it from the rolling.
A steel shortage is the result. Thus, the chemistry must also be produced
reliably.

Surface defects are readily found by visual inspection on the cold billets.
Frequently these defects are not visible when the billet is hot. Many surface
defects roll into very tight defects on the rolled product and can only be
found by M.P.I. or Eddy Current testing. Even the visible surface defects

require the handling of at least five times more bars than billets to find the defective ones. This can lead to extreme bottlenecks at the finishing end of the mill.

Internal defects can only be detected by deep etching or sulphur printing a billet or product slice, or by ultrasonic testing of the rolled product. Billet slices cannot usually be processed quickly enough to remove defective materials from the rolling.

Successful inspection and rejection when done hot and in-line at the billet stage will result in an imbalance in the casting and rolling rates. This is particularly significant considering the fact that continuous casters usually produce many defective billets during a specific production problem even though monthly defect levels are very low. Thus it becomes a necessity to have "defect free" billets for hot charging.

The question of "defect free" steel must be addressed at this point. It is important to recognize that a defect is something that prevents the product from satisfying the customer's requirements and should not be considered as something which prevents the billet from being technically perfect.

The process must produce billets of acceptable quality as well as producing them at the correct rate. In other words, the process must be statistically capable. This can only be accomplished through process control not through inspection.

Implementation

The growth of hot charging should have two phases in an existing facility. The initial phase should involve a product for which the caster can produce "defect free" billets. This allows the plant to assess the physical and economic practicality of the technique without the constraints of billet quality. This appears to be a common phase in the move towards realization of the cost reductions and it is one Lasco has completed.

The second phase is to assess the equipment and techniques necessary to produce "defect free" steel for the other products in the plant. We are presently involved in this longer term phase at Lasco. It involves improvements in the level and variability of billet quality which may reduce costs for more than hot charging. The production of defect free billets for hot charging results in no downgraded billets, no billet conditioning and scheduling benefits even if the billets are cold charged.

There is usually more than one solution to a problem and the trick is to find the one which results in the highest payback. Given our present shop operating constraints at Lasco it was decided that the best way to improve quality capability was through the implementation of O.P.C. (Operator Process Control), installation of a L.A.R. (ladle arc refiner) and E.M.S. (ElectroMagnetic Stirring).

O.P.C. is Lasco's version of S.P.C. (Statistical Process Control). We have replaced Statistical with Operator to stress the fact the operator is the one who really controls the process and needs the statistical tools to evaluate control. As stated earlier, the control of the process of billet making is the only way to achieve "defect free" billets. By instituting O.P.C. it is possible to address all aspects of the operation and achieve statistical capability in the process. This allows us to be repeatable in such critical quality aspects as temperature, chemistry, deoxidation, reoxidation control, slag control at tap, tundish depth and casting speed. Control of variability in all of these areas is most important to the consistency of billet quality.

Repeatability of chemical analysis and temperature were considered to be such critical aspects that a L.A.R. was built at Lasco. The L.A.R. allows for electric arc heating of the steel in the ladle while it is being electromagnetically stirred under an inert slag. This ensures temperature and chemistry control thus preventing any reoxidations. The L.A.R. process has also benefited from the O.P.C. approach to controlling a process.

Once consistency is achieved then decisions can be made on new or modified equipment that will allow the process to improve the base quality level.

The decision made at Lasco was to add E.M.S. in the mould and at the final solidification point. This equipment was expected to be most effective in decreasing our process variability and in increasing our base quality level thus allowing the consideration of additional products for hot charging.

The results of these process control programs on the ability to produce "defect free" billets is shown in Fig. 5. The large improvement at the beginning of 1988 can not be credited to E.M.S. alone but rather continued refinement through O.P.C. in conjunction with E.M.S.

DECREASE IN NON-PRIME BILLETS AS THE RESULT OF PROCESS CONTROLS

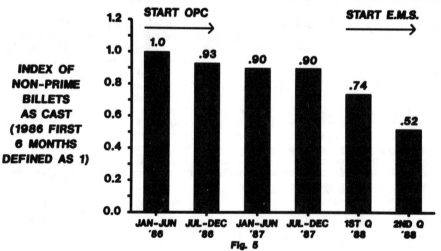

Fig. 5

SUMMARY

Significant cost reduction can be obtained by the use of hot charging. However, existing plant facilities and product mix limit the realization of 100% of the cost reductions.

Hot inspection is not an effective method for successful implementation of hot charging. It is necessary to produce "defect free" steel to be most successful in a hot charging program and process control is essential to achieve "defect free" steels.

Perhaps the most important benefit of hot charging has not yet been discussed.
Hot charging is a readily visible economic benefit that drives the search
toward improved quality or "defect free" steels.   "Defect free" steels are
necessary to enable hot charging.   However, the cost reductions gained through
no defects can be realized even when cold charging and these reductions are
greater than those associated with hot charging.

BIBLIOGRAPHY

Lu, W.K. (May 1985). Requirements of Hot Charging of Continuously Cast Products.
McMaster University Symposium No. 13.

MEASUREMENT OF OSCILLATION MARKS ON CONTINUOUSLY
CAST BILLETS WITH AN AUTOMATED PROFILOMETER

I.A. Bakshi, E. Osinski, I.V. Samarsekera and J.K. Brimacombe

The Centre for Metallurgical Process Engineering
The University of British Columbia
Vancouver, B.C. V6T 1W5
Canada

ABSTRACT

Oscillation marks on billets have been linked to the formation of off-squareness, off-corner internal cracks, transverse cracks, bleeds in the mould and, in severe cases, break-outs below the mould. Knowledge of oscillation mark characteristics thus is essential to the achievement of good billet quality. To facilitate the characterization of oscillation marks, a device has been developed which automatically measures the profile of a billet surface. The depth and pitch of the oscillation marks can be determined simultaneously, with the data being fed to to a computer for storage and analysis. A key feature in the design has been the incorporation of a "mathematical wheel", developed through computer analysis of previous billet measurements, to filter out surface variations between oscillation marks. This wheel allows a travelling surface profile to be incorporated automatically into the oscillation measurements which greatly simplifies interpretation of the results. To illustrate the utility of the automated profilometer, measurements were made on a billet cast through a machine having a high breakout frequency. The profilometer revealed that oscillation marks on opposite faces differed by a factor of two, likely as a result of poor lubricating oil distribution.

KEYWORDS

Billet, Continuous Casting, Measurement, Mould, Profilometer, Oscillation Marks, Quality, Steel, Surface, Topography.

INTRODUCTION

Steelmakers recently have had to grapple with increasingly stringent standards of surface quality applied to continuously cast billets, notably destined for markets such as forging products. Previously, oscillation marks were not thought to be a problem unless they were so deep as to cause a lap in the finished product. However it has been shown that oscillation marks have a profound influence on local heat transfer and solidification (Samarasekera, Brimacombe and Bommaraju, 1984) as they locally increase the mould/shell gap width and hence reduce the heat-extraction rate. This results in a locally thin shell and possibly cracking at the base of the oscillation mark which, in severe cases, can cause a break-out below the mould. Another possible source of poor surface quality is inadequate or uneven mould lubrication which can lead to sticking in the mould, with accompanying tears, bleeds and other surface problems.

In spite of the importance of the billet surface to the achievement of a good product and the easy accessibility of the surface for inspection, not much work has been done to analyze billet topography. This has no doubt been due to the difficulty of interpreting the billet surface, as well as the tedious and time consuming nature of the work.

In order to study the influence of mould design, oscillation characteristics and operating parameters on billet quality, a means of measuring the surface topography of billets easily and accurately was needed. Previously, a somewhat crude modified welding machine running on a track had been employed (Brimacombe and co-workers, 1984) for this purpose. A linear variable displacement transducer (LVDT) was affixed to the machine and its calibrated electrical signal yielded the vertical distance between it and the billet surface. A critical problem with this apparatus was that chart traces of the billet surface profile were obtained only relative to the moving machine. This required that the charts be inspected visually and hand measurement made to determine the depth of each oscillation mark. The best accuracy that could be achieved with this technique was ± 10% of the oscillation mark depth.

The goal of this study therefore, was to design an automated billet profilometer that would simplify the process of measurement and yield surface topography data in digital form. This would allow for easy mathematical analysis of the data, hopefully yielding meaningful results in a fraction of the previous time.

MATHEMATICAL WHEEL CONCEPT

As described previously, measurements made with the simple motor-driven LVDT which determined vertical differences between its fixed trajectory and the billet surface had numerous problems. Among these were:

1) The billet surface was not precisely parallel to the trajectory of the moving LVDT. This resulted in the recorder trace either continually increasing or decreasing as the LVDT travelled along the billet length.
2) In addition the billet surface was wavy or undulating in nature. These low frequency undulations are not connected with the oscillation marks as they occur over much longer distances. However, the LVDT trace records these waves making the determination of oscillation mark depths very subjective.
3) Problems were encountered with warping of the track upon which the LVDT rides. This distortion would be inaccurately recorded as a variation of the billet surface.

All of the above effects can result in displacement readings greater than that of the oscillation marks themselves. Figure 1 shows an example of a poor trace obtained with the old equipment. As can be seen, the surface varies in position making it very difficult to decipher the actual oscillation mark depths.

To analyse data in the face of these problems, a mathematical algorithm was sought which would remove all unwanted outside disturbances and leave only a record of the true oscillation marks. Specifically, any undulating variations in billet surface of a non-periodic or long period nature or (in terms of Fourier spectrum analysis) all of the frequencies outside the range of interest should be filtered. One solution is to process the data using a mathematical filter based on the concept of a moving wheel. The wheel is chosen large enough in diameter so as to not enter the periodic depressions of the oscillation marks, while still entering

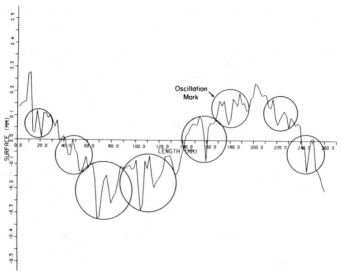

Fig. 1. Billet surface profile obtained with the first
generation measurement equipment.

all the depressions having a longer wavelength as it rolls along the billet
surface. The measurement probe is considered to be attached to the center axis of
the wheel recording the vertical distance between the rim of the wheel and the
surface of the billet. The obtained "filtered" probe record is then equal to the
distance between the envelope of wheels "rolled" and the measured billet surface.
More specifically the algorithm developed was as follows:

1) The length of the sample is discretized into small increments, creating a
   one-dimensional grid.
2) The record of rough data from the LVDT probe is input into the grid and is
   considered to be the billet "surface".
3) For each point along the grid, the lowest position of the circle is calculated
   as the point where the circumference is just above the surface. This is
   accomplished through an iterative procedure, in which the middle of the circle
   is lowered and the position of its circumference is checked relative to the
   surface at each increment.
4) Once having a set (body) of these circles, their envelope is calculated as a
   locus of the points having the lowest position for each point of discretized
   length. This envelope is a reference for calculation of the "filtered"
   surface, which is computed as the difference between the original surface and
   the envelope.

This procedure is shown in Fig. 2. Curve 1 is the unprocessed surface of the
billet; Curve 2 is the envelope of the circles being "rolled" on the surface of
the billet and Curve 3 is the resultant "filtered" surface. The determination of
the proper wheel radius to be used in the analysis was somewhat arbitrary. The
optimum value depends on the frequency of the oscillation marks and overall
roughness of the billet surface which varies primarily with carbon content. Too
large a radius would allow the algorithm to filter out only part of the unwanted
frequencies, while too small a radius would result in filtering out part of the
signal of interest (the oscillation marks). Utilizing different radii of wheels
for each case, while feasible mathematically, would be impractical in an actual
machine. Fortunately an analysis of the mathematical results showed that the

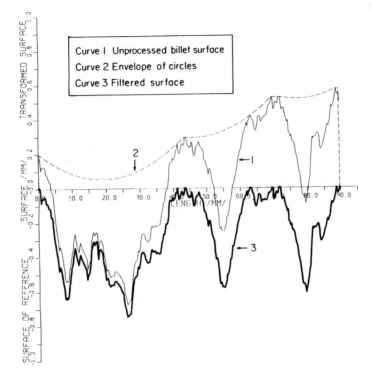

Fig. 2. The procedure of mathematically filtering a billet
surface profile.

optimum value falls within the range 0.5 to 3.0 meters, and that above a value of
0.5 meters, the influence of wheel radius on the filtration process is not
significant.

The processed billet surfaces obtained by the preceding algorithm proved to be
easily readable and highly amenable to further statistical processing. From this
point of view, the algorithm proved satisfactory. However the calculations were
costly, and the acquisition of the rough data using the old apparatus was still
difficult and time consuming. The decision was made to build measurement
equipment incorporating the mathematical wheel filtering concept.

NEW BILLET PROFILOMETER

Building a machine to measure billet topography utilizing this mathematical wheel
concept quickly introduced some practical limitations. Knowing that the diameter
of the wheel was supposed to be one-half meter or greater, it was quickly
established that an actual wheel of this diameter supporting an LVDT at its center
was impractical. It was decided instead to modify the wheel concept somewhat and
employ only a small segment of the wheel circumference. The wheel segment would
slide rather than roll along the billet surface. The equivalent of a one meter
radius wheel was used for the machine; and the LVDT probe, instead of being fixed
to the wheel axis, was fastened directly to the segment circumference. The
arrangement of wheel, LVDT and holder can be seen schematically in Figs. 3 and 4.

Figure 3 shows the arrangement of the pivoting arm assembly which allows the wheel segment to move up and down as it travels along the billet surface while Fig. 4 illustrates the wheel segment and LVDT holder design in more detail. A special low friction, highly wear resistant, ultra-high molecular weight polyethylene was used for the wheel segment.

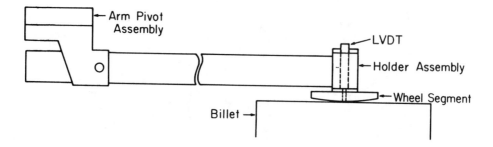

Fig. 3.   The   automated   profilometer   pivoting   arm arrangement.

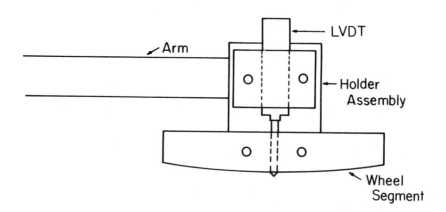

Fig. 4.   The automated profilometer wheel segment and LVDT holder arrangement.

The profilometer that was constructed is pictured in Fig. 5. The final design consisted of three independent arms and LVDT probe assemblies so that measurements could be made at the billet center and two off-corner sites simultaneously.

The wheel segments and supporting arms remain stationary with the billet moving instead. The billet is carried on a heavy cast iron table driven by a worm gear capable of multiple speeds. The table has a total travel distance of one meter, making the machine suitable for long billet length measurements or adaptation for mould measurements, if desired at a later date.

The LVDT's used are the spring-loaded AC type, capable of extremely high accuracy (> 10 μm). A fine point carbide tip forms the probe end in contact with the billet surface to ensure that the probe follows the billet contours exactly. The

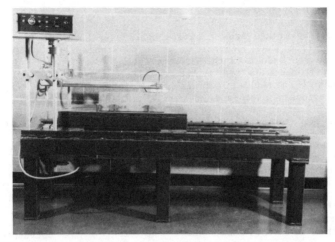

Fig. 5. Photograph of the completed billet profilometer.

LVDT is electrically connected to a signal conditioner which generates a 0 to 5 Volt DC signal proportional to the displacement of the LVDT. This signal is then fed into a Data Translations analog-to-digital converter board installed in a Compaq portable personal computer. There, the analog voltage signal is converted into a digital format, scaled into a numerical displacement and finally stored in a table. This data collection process was greatly simplified through the use of Labtech Notebook, a general purpose laboratory data acquisition program.

The numerical data which then is obtained can be easily manipulated statistically. Currently Lotus 123 is being utilized to perform statistical calculations as well to generate graphs of the surface topographies. A modified variance function has been favoured to best represent the relative roughness of the billet surfaces as it highlights the variation in billet topography by taking the square of the differences from the null line, thereby emphasizing the degree of surface roughness.

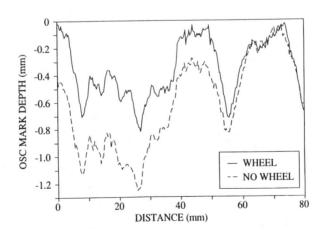

Fig. 6. Billet surface profiles made with and without the travelling wheel.

RESULTS AND DISCUSSION

Two surface profiles taken from a sample billet are shown in Fig. 6. One trace was made without the wheel segment in order to simulate the older machine while the second trace was made with the new travelling wheel design. There is clearly a significant difference in the billet profiles generated by the two techniques. This is caused by the billet surface undulations influencing the displacement readings as previously discussed. Comparing the surface profiles generated by mathematically filtering the data from the non-wheel run shown in Fig. 2 to that of the profile produced by using the physical wheel shown in Fig. 6, the two can be seen to be in close agreement. A further calculation to verify this point was made using Fourier analysis on the three waveforms: one from the data obtained without using the wheel (Fig. 7), one from mathematically processing this data (Fig. 8) and one obtained from the physical wheel (Fig. 9). Figure 7 shows the existence of a strong peak on the power spectrum corresponding to frequencies lower than that of oscillation marks. As can be seen from Figs. 8 and 9, application of either the physical or mathematical wheel, results in nearly identical Fourier spectra. Both techniques have removed nearly all the unwanted frequencies from the raw data, while the single peak corresponding to the oscillation mark frequency remains predominant.

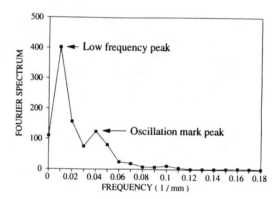

Fig. 7.   Fourier analysis of billet trace obtained without using the travelling wheel.

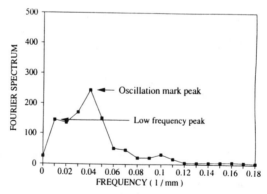

Fig. 8.   Fourier analysis of billet trace obtained without using the travelling wheel but processed mathematically.

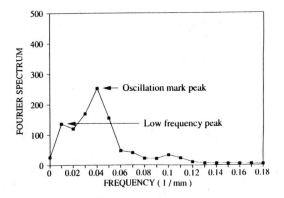

Fig. 9.   Fourier analysis of billet trace made using the
          travelling wheel segment on the new profilometer.

To illustrate the usefulness of making billet topography measurements, a series of
tests were made on billets cast through a machine having a high break-out
frequency. The billet chemistry and casting conditions are given in Tables 1 and
2.   The billet surfaces were sandblasted then measured with the new automated
profilometer utilizing the travelling wheel segment.   The tests revealed marked
differences in the surface roughness on the opposite faces of the billets. Figure
10 shows the surface profiles of the same billet taken on the two faces untouched
by the pinch rolls.   The average depth and variance of the surface measurements
made are presented in Table 3.   There is clearly a significant difference in the
surface roughness between the two sides.   The measured average oscillation mark
depth differed by a factor of two while the difference in variance was almost four
times higher on the one side than the other.

TABLE 1   Billet Chemistry

C	Mn	P	S	Si	Cu	Cr	Al
0.29	1.41	0.024	0.029	0.32	0.27	0.14	0.013

TABLE 2   Casting Conditions

Stroke Length (mm)	Osc. Frequency (#/min)	Casting Speed (m/min)	Negative Strip Time (sec)	Mould Water Velocity (m/s)	Tundish Superheat (°C)
19.0	100	2.67	0.21	8.0	70

This and other billet surface measurements showed a dramatic and consistent
difference in surface roughness between the two unrolled sides of the billets cast
on this machine. This information, in addition to measured heat flux profiles,
led to the conclusion that sticking of the billet had taken place in the mould,
possibly indicating a problem with uneven mould lubrication (Samarasekera and
Brimacombe, unpublished work).

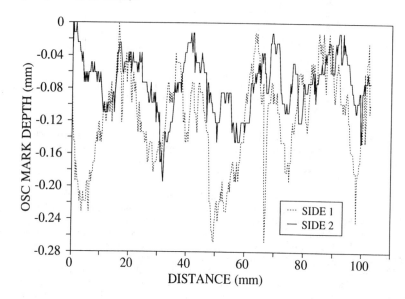

Fig. 10. The billet surface profiles measured on two
         opposite faces.

TABLE 3 Billet Surface Measurements

CALCULATION	SIDE 1*			SIDE 2*		
	A	C	B	A	C	B
AVERAGE DEPTH (mm)	0.145	0.243	0.145	0.072	0.131	0.077
VARIANCE (mm)^2	0.0259	0.0676	0.0256	0.0065	0.0198	0.0076

*Side 1

*Side 2

CONCLUSION

An automated profilometer for measuring the surface topography of continuous cast steel billets has been constructed. The machine design featured the incorporation of a travelling wheel segment to filter out the surface fluctuations between oscillation marks. This allowed a travelling surface profile to be taken into account automatically when making oscillation mark depth measurements, which greatly simplified the interpretation of the results. Measurements made on billets from a machine having a high break-out frequency revealed that the oscillation marks on opposite billet faces differed by a factor of two, likely as a result of poor lubricating oil distribution.

ACKNOWLEDGEMENT

The authors wish to thank Geoff Webb of Sciema Technical Services for his invaluable assistance in the final design and building of the automated billet profilometer. The financial support of the Natural Sciences and Engineering Research Council of Canada and the following companies – Courtice Steel, Hatch Associates, Ivaco, Sidbec-Dosco, Slater Steels, Stelco and Western Canada Steel – is greatly appreciated.

REFERENCES

Brimacombe, J.K., I.V. Samarasekera, N. Walker, I. Bakshi, R. Bommaraju, F. Weinberg, and E.B. Hawbolt (1984). Mould behaviour and solidification in the continuous casting of steel billets. I. Industrial trials. ISS Trans., Vol. 5, 71-77.

Labtech Notebook, Laboratory Technologies Corp., 255 Ballardvale Street, Wilmington, MA. 01887.

Lotus 123, Lotus Development Corp., 55 Cambridge Parkway, Cambridge, MA. 02142.

Samarasekera, I.V. and J.K. Brimacombe. Unpublished work, The University of British Columbia.

Samarasekera, I.V., J.K. Brimacombe, and R. Bommaraju (1984). Mould behaviour and solidification in the continuous casting of steel billets. II. Mould heat extraction, mould-shell interaction and oscillation mark formation. ISS Trans., Vol. 5, 79-94.

HEAT EXTRACTION CAPABILITY OF CONTINUOUS-CASTING BILLET MOULDS

I.V. Samarasekera and J.K. Brimacombe

The Centre for Metallurgical Process Engineering
The University of British Columbia
Vancouver, B.C. V6T 1W5
Canada

ABSTRACT

Heat-flux profiles determined from wall temperature measurements on instrumented single-taper, double-taper and specially designed constrained moulds are presented. An application of the conventional resistance analogue to the measured mould heat-transfer rates has revealed the significant role of conduction through the solidifying shell particularly in the lower part of the mould and at low casting speeds, in limiting overall heat transfer from the solidification front to the mould cooling water. Thus, it has been shown that the mould can be divided into two zones: an upper region of gap-resistance dominance in which heat extraction can be influenced by factors altering the gap width like taper and distortion; and a lower region in which the shell resistance is significant. For a 0.30-pct. carbon steel, the shell resistance first becomes comparable to the gap resistance at a shell thickness of 6 mm which corresponds to a time (distance below the meniscus/casting speed) of 7.8 s. To permit determination of the influence of casting speed on mould heat transfer, a relationship linking heat transfer and time in the mould, which includes the resistances of the shell, gap, mould wall and mould cooling water, has been formulated for a 0.30-pct. carbon steel. The surface temperature, shell thickness, and billet shrinkage profiles have been determined for the same steel with the aid of a mathematical model for two casting speeds. It has been shown that for low casting speeds, when employing double-taper moulds, the second taper must be markedly reduced as compared with the requirement for high casting speeds because heat extraction lower in the mould is reduced due to the increased conduction resistance of the thicker shell. This is contrary to current practice where both upper- and lower-zone taper in a double-taper mould are increased for large section sizes cast at low speeds. The implication of these findings with respect to the formation of transverse depressions and transverse cracks has been explored.

KEYWORDS

Heat transfer; mould taper; shell growth; transverse cracking.

INTRODUCTION

Control of mould heat transfer and interaction between the mould and solidifying shell is central to the continuous casting of quality steel billets at high production rates. However, the heat extraction is complex varying in magnitude from mould corner to midface as well as from meniscus to mould bottom due, in large part, to the formation of a gap between the steel surface and the copper wall. Paricularly in the upper part of the mould, conduction across the gap is the greatest resistance to mould heat extraction and is, therefore, rate limiting. Consequently, factors which influence the width of the gap such as shrinkage or bulging of the solidifying shell, oscillation-mark depth, rough shell surface (in low-carbon grades), mould taper and mould distortion, also affect mould heat extraction. Moreover the composition of the material in the gap - gaseous oil pyrolysis products or mould flux - alters heat transfer by changing the gap conductivity.

Owing to the complexity of gap behaviour, an understanding of heat transfer in the mould and its effect on billet quality, has been sought by mounting a campaign of plant trials over the past decade in steel plants throughout North America (Bommaraju, 1984; Bommaraju, 1988; Brimacombe, 1984; Pugh, 1988; Samarasekera, 1988). Typically in the tests, a mould tube has been instrumented with an array of wall-embedded thermocouples which have yielded axial temperature profiles within the copper wall. The temperature data has been converted to axial heat-extraction profiles with the aid of a mathematical model (Samarasekera, 1979, 1982) and knowledge of the cooling water velocity which is measured in situ with pitot tubes (Berryman, 1988). For each heat in which mould temperatures have been measured, billet samples also were cut for subsequent examination with respect to oscillation mark depth, internal structure, shape, cracks and solidification bands. In this way, links between mould heat extraction/thermomechanical behaviour and billet quality were established which have led to improvements in mould design and operation (Brimacombe, 1986). These changes have resulted in a reduction of off-squareness, off-corner internal cracks and break-outs.

In this paper, mould heat-transfer measurements are presented from plant trials involving single- and double-taper moulds as well as a mould of special design to prevent mould tube distortion. The effect of carbon content of the steel also is included. An examination of the measured heat-flux profiles has revealed the significant role of conduction through the solidifying shell, particularly in the lower part of the mould and at low casting speeds, in limiting overall heat transfer from the solidification front to the mould cooling water. As shall be seen, this finding is important for the design of mould tapers and has implications for the prevention of transverse cracks and depressions. Because an understanding of the effect of mould taper and casting speed on heat extraction depends on an appreciation of the magnitude of the shell conduction resistance, the first section of the paper is devoted to a classical resistance analysis of the mould heat transfer.

RESISTANCES TO MOULD HEAT EXTRACTION

In the mould, superheat and latent heat released at the solidification front must be transferred to the cooling water by several paths in series, as shown in Fig. 1:

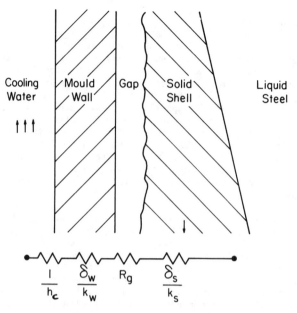

Fig. 1.   Schematic diagram showing paths of heat
          transfer from molten steel to cooling water in
          the mould.   Series resistance analgue also
          shown.

[i]   conduction through the solidifying shell;
[ii]   conduction and (to a lesser extent); radiation across the steel/mould
       gap;
[iii] conduction through the copper mould wall; and
[iv]   convection to the mould cooling water.

Under conditions of steady state, one-dimensional heat flow (time invariant,
linear temperature gradients through the shell, gap and mould wall), which hold
approximately in continuous casting, the heat flux from the solidification
front at the solidus temperature T_s to the cooling water at temperature T_w can
be written as follows:

$$q = \frac{1}{R_T} (T_s - T_w) \tag{1}$$

where R_T is the total resistance to heat flow given by

$$R_T = \frac{\delta_s}{k_s} + R_g + \frac{\delta_w}{k_w} + \frac{1}{h_c} \tag{2}$$

(shell) (gap) (wall) (cooling water)

The heat flow described by Eqs.(1) and (2) is analogous to current flow through
a series resistance circuit connecting different electrical potentials (Kreith,

1980), as shown in Fig. 1. Typically the thermal resistances of the mould wall and cooling water are small, of the order of $2.6(10^{-2})$ and $4.0(10^{-2})$ m^2 °C/kW respectively and often can be neglected.

To assess the relative importance of the remaining shell and gap resistances, heat fluxes measured in a double-taper mould during the casting of 0.19-, 0.33- and 0.45-pct. carbon steels, as shown in Fig. 2, were analysed initially. For each steel grade, the total thermal resistance was calculated from Eq.(1). The solid shell resistance then was computed by first predicting the shell thickness profile using a finite-difference model (Brimacombe, 1976), with the surface boundary condition characterized by the measured heat fluxes for each steel grade. The calculated shell profiles for the 0.31- and 0.45-pct. carbon steels cast at 33-34 mm/s are shown in Fig. 3. The computed shell thickness at any level subsequently was divided by the steel thermal conductivity to yield the shell resistance. Finally, the gap resistance, R_g, was calculated by rearranging Eq.(2).

$$R_g = R_T - (\frac{\delta_s}{k_s} + \frac{\delta_w}{k_w} + \frac{1}{h_c}) \qquad (3)$$

Axial profiles of the gap and shell resistances, calculated in this manner, are shown in Figs. 4 and 5 respectively for the 0.31- and 0.45-pct. carbon steels. Thus it can be seen that the gap resistance dominates for about 250 mm below the meniscus in both cases but lower in the mould, the two resistances are comparable. The shell thickness at which the resistances become similar is about 6 mm, Fig. 3.

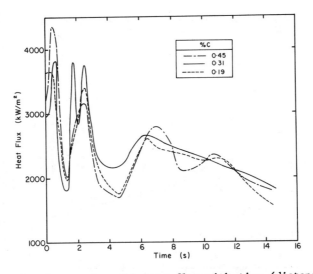

Fig. 2. Change of mould heat flux with time (distance below meniscus/casting speed) in a double-taper mould. Measured during the casting of 0.19-, 0.31- and 0.45-pct. carbon steels at 33-34 mm/s.

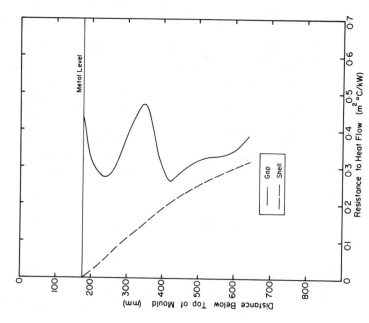

Fig. 4. Axial profiles of gap and shell thermal resistances determined for the casting of a 0.31-pct. carbon steel through a double-taper mould at 33–34 mm/s.

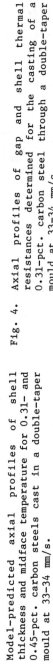

Fig. 3. Model-predicted axial profiles of shell thickness and midface temperature for 0.31- and 0.45-pct. carbon steels cast in a double-taper mould at 33–34 mm/s.

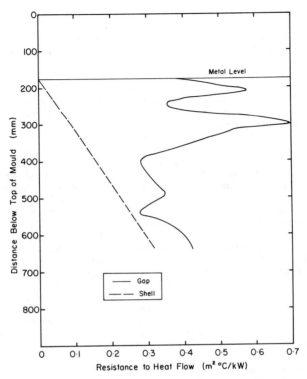

Fig. 5. Axial profiles of gap and shell thermal
resistances determined for the casting of a
0.45-pct. carbon steel through a double-taper
mould at 33-34 mm/s.

Thus the mould in question can be divided roughly into two zones: an upper
region of gap-resistance dominance in which heat extraction can be influenced
by factors altering the gap width like taper and mould distortion; and a lower
region in which the shell resistance is significant so that the effect of the
gap resistance is diminished. In the lower zone, the influence of mould taper
on heat extraction concomitantly will be less. If the shell thickness of 6 mm
can be taken as a rough critical value for a significant shell resistance, then
the length of the upper zone will vary linearly with casting speed. The taper
applied in the lower part of the mould must reflect the importance of the shell
resistance; but this has not been the case in earlier work (Dippenaar, 1986).
The question of determining mould tapers for different casting speeds will be
taken up again later in the paper.

The division of the mould into two thermal zones also can be seen from the
axial profiles of midface temperatures shown in Fig. 3. From the meniscus to
about the 425-mm level, the surface temperature decreases rapidly from the
pouring temperature to about 1050°C. Below this point, the temperature decline
continues to diminish to reach an exit value of 850°C. Thus the lower part of
the mould acts more as a zone of thermal steady state with the rate of heat
conduction through the shell roughly matching heat flow across the gap.

MEASURED MOULD HEAT EXTRACTION

Further insight on the mechanism of heat extraction in the mould may be gained by comparing axial heat-extraction profiles from the double-taper mould, Fig. 2, to measurements made on two other mould systems. Details of the three systems are given in Table 1. Figure 6 shows heat-extraction profiles at the straight wall of a single-taper, curved mould for 0.13- and 0.30-pct. carbon steels (Grill, 1976). These heat-extraction profiles have been determined with the aid of a mathematical model from time-averaged mould temperatures measured during operation (Samarasekera, 1984). It is seen that the heat-extraction rates are lower for the 0.13-pct. carbon steel than for the 0.30-pct. carbon steel, an effect that has been well documented in earlier publications (Grill, 1976; Hurtuk, 1982; Singh, 1976; Wolf, 1980). The differences in the heat-extraction rates have been attributed to the greater wrinkling of the solidifying shell, creating larger gaps, in low-carbon steels close to the meniscus (Grill, 1976). This is thought to be caused by shrinkage associated with the δ-γ transformation which in 0.1-pct. carbon steels takes place solely in the solid-state (relative to higher carbon steels where the peritectic reaction involving liquid is involved). It is important to note that the difference in heat extraction between the two grades diminishes with distance down the mould, the effect becoming insignificant at approximately 8.5 s which translates to a distance a 455 mm at a casting speed of 33 mm/s. This can be attributed to a reduction in the importance of the gap resistance in controlling heat flow in the lower region of the mould, as discussed in the previous section. From Fig. 7, which shows the variation of the gap and shell resistances down the length of the mould for a medium-carbon steel, it is evident that the shell resistance becomes significant, relative to the gap resistance, below ~ 425 mm. For the low-carbon steel, Fig. 8, the shell resistance is a smaller fraction of the overall resistance as compared with the medium-carbon steel presumably due to the greater wrinkling of the surface of these billets. Nevertheless, in the lower mould, the sum of the shell and gap resistances for the two grades of steel becomes very similar resulting in virtually identical heat-extraction rates; and the influence of modest changes in gap width on overall heat transfer is markedly reduced.

From this discussion it is evident that mould shape, as affected by either distortion or taper, will have its greatest influence on heat-extraction rates in the upper region of the mould. This is clearly demonstrated in Fig. 9 which compares the axial heat-extraction profiles for a 0.30-pct. carbon steel cast through the single-taper and double-taper moulds as well as on a fully constrained mould (the latter was designed to prevent outward bulging of the mould due to differential thermal expansion). Details of each mould are given in Table 1. In the upper mould region the heat-extraction profiles of both the single- and double-taper conventional mould tubes exhibit several maxima and minima which are noticeably absent in the heat-extraction profile for the fully constrained mould. In the latter the copper tube was prevented from expansion in a transverse plane by a specially designed constraint system. Thus it may be argued that with the conventional mould systems which are unrestrained over their working length, the outward bulging of the tube (Samarasekera, 1982) below the meniscus is responsible for the local increase in gap width and a decrease in the heat-extraction rate. The heat flux exhibits a local minimum at the point of maximum outward bulging, which occurs at 1 s in the single-taper tube and 1.3 s in the double-taper tube. The second maximum in the heat-extraction profiles, which occurs around 1.8 s for both the single- and double-taper tubes, could be attributed to the closing of the gap caused by the inward movement of the mould wall below the position of the peak outward bulge. With the double-taper tube, a third peak is seen in the heat-flux profile. This may be related to the fact that the mould tube was rectangular casting 127 x 178 mm^2 billets. Then because the cross-sectional dimensions are

TABLE 1   Design and Operating Details of Mould Systems Employed in Plant Trials

Parameters	Single Taper	Double Taper	Constrained Mould
Mould Wall Thickness (mm)	7.94	12.7	18.72
Mould Length (mm)	812.8	812.8	812.8
Section Size (mm^2)	144 x 144	127 x 171.5	144 x 144
Taper %/m	0.6	2.6 (0-330 mm) 0.48 (330-812 mm)	0.6
Type of Support	Keeper plates into slots on two straight sides.	Keeper plates into slots on four sides.	Lateral expansion prevented. Longitudinal expansion allowed.
Casting Speed	33 ~ 34 mm/s	34 ~ 37 mm/s	32 mm/s
Metal Level (mm)	175	115	152

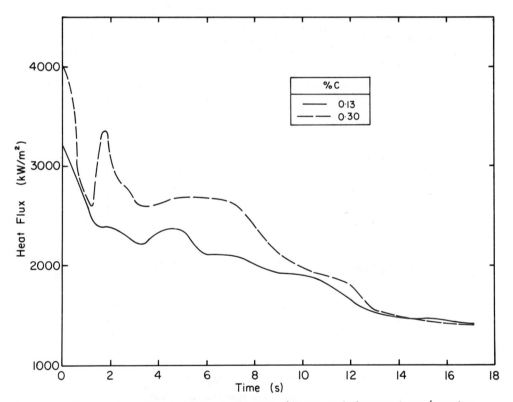

Fig. 6.   Change of mould heat flux with time (distance below meniscus/casting speed) in a single-taper mould. Measured during the casting of 0.13- and 0.30-pct. carbon steels at 33 ~ 34 mm/s.

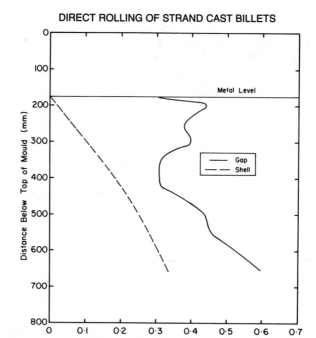

Fig. 7. Axial profiles of gap and shell thermal resistances determined for the
casting of a 0.30-pct. carbon steel through a single-taper mould at
33 ∿ 34 mm/s.

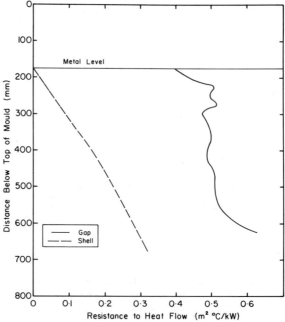

Fig. 8. Axial profiles of gap and shell thermal resistances determined for the
casting of a 0.13-pct. carbon steel through a single-taper mould at
33 ∿ 34 mm/s.

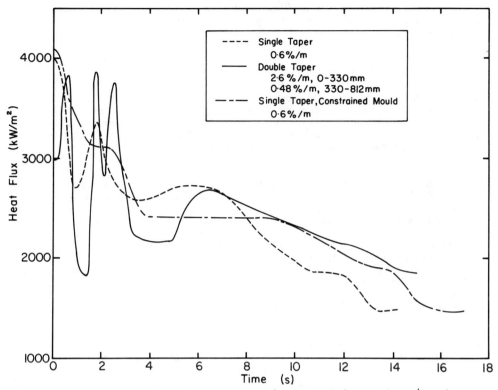

Fig. 9.   Change of mould heat flux with time (distance below meniscus/casting
          speed) in a single-taper, double-taper and constrained mould measured
          during the casting of 0.30-pct. carbon steel.

unequal, the peak distortion occurs at different depths below the top of the
mould on adjacent faces. The outward movement of one face is accompanied by an
inward movement of the adjacent faces owing to rotation of the corners that are
cold and rigid; the result is a rebound in heat flux on the adjacent faces.
This phenomenon also occurs when the constraints on adjacent faces at the top
of the mould are different as in the case of mould tubes that have slots for
retainer plates only on the two straight walls (Samarasekera, 1984). The lower
heat-extraction rate at the meniscus with the double-taper mould may be
attributed to reduced mechanical interaction between the mould and the shell
during the negative-strip period of the oscillation cycle (Brimacombe, 1986) as
a result of the steeper positive taper. The slight increase in heat transfer
seen in Fig. 9 over ~ 6.5 s for the single- and double-taper moulds is probably
due to outward bulging of the shell driven by ferrostatic pressure. With the
double-taper mould, this occurs just below the point of taper change. Below 7
s the heat-extraction rates along the midface of the three mould systems are
very similar decreasing more severely for the single-taper mould. As was shown
earlier from Figs. 4 and 7, the resistance of the shell is comparable to the
resistance of the air gap over this region and the dependence of the
heat-extraction rate on minor changes in gap width is reduced. The
heat-extraction rates are also significantly lower than over the first 3 s of
contact with the mould owing to the resistance to heat flow imposed by the
growing shell. Thus it may be concluded that, irrespective of design, the
mould can be divided into two zones as described earlier: an upper zone in

which the gap resistance dominates rendering the heat transfer sensitive to mould design and steel composition, and a lower region where the shell resistance has a significant effect.

SHELL GROWTH AND MOULD HEAT TRANSFER

Since the heat-extraction rate over the lower regions of the mould is strongly dependent on the shell thickness, it is important to examine the shell growth for a variety of conditions. This is particularly important to assess the effect of changes in casting speed on the heat-flux profile which, in turn, will influence the billet surface temperature in the mould and the required taper.

Figure 10 shows the computed shell thickness and surface temperature profiles of the 0.13-pct. and 0.30-pct. carbon steels during casting at a speed of 32 mm/s through a 144-mm square, single-taper mould. The shell thickness at which the resistance to heat flow through the gap and the shell first becomes comparable for the 0.30-pct. carbon steel is ~ 5.8 mm which is very similar to the value of 6.0 mm obtained with the double-taper mould in Fig. 3. For both the medium- and low-carbon steels, it is also seen that the surface temperature decreases rapidly from the meniscus level down to a depth of ~ 400-425 mm; below this level the surface temperature changes only gradually. This zone of thermal steady state was observed also in the double-taper mould, Fig. 3, although in that case cooling was slightly greater.

Figure 11 shows the shell thickness profiles for 0.30-pct. carbon steel computed with the aid of the finite-difference model (Brimacombe, 1976) using the heat-flux profiles determined for the single-taper, double-taper and constrained moulds. To permit comparison, the computations have been made for a 144-mm square section being cast at 32 mm/s. The shell thickness at any position below the meniscus is highest with the constrained mould but toward the bottom there is little difference among the three cases. At the mould exit, the shell thickness is 12.0, 12.4 and 12.5 mm with the single-taper, double-taper and constrained tubes respectively.

The shell-thickness profile for the 0.30-pct. carbon steel with the double-taper mould has been fitted with a power law relationship to time. Thus the shell growth can be characterized by the following equations:

$0 < t < 6.0$ s

$$\delta_s = 0.64 \ t \tag{4}$$

$6.0 < t < 11.0$ s

$$\delta_s = 0.905 \ t^{0.904} \tag{5}$$

$t > 11.0$ s

$$\delta_s = 1.46 \ t^{0.706} \tag{6}$$

In this work, it has been established that the resistance of the shell becomes comparable to that of the gap when the shell is approximately 6.0 mm thick; this corresponds to a time of 7.8 s. For times less than 7.8 s, the heat extraction in the mould is largely controlled by the gap. The heat flux –

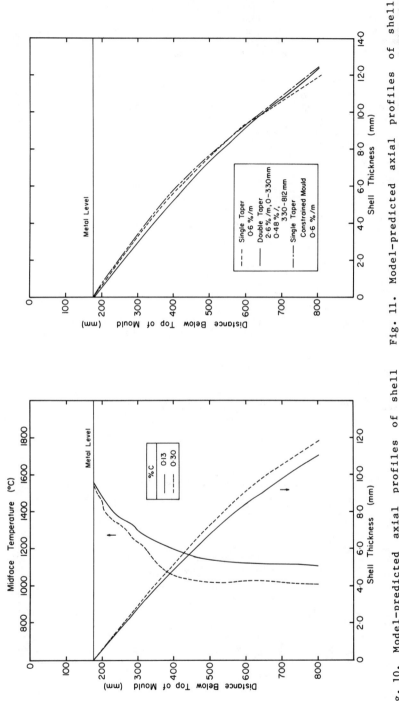

Fig. 11.   Model-predicted axial profiles of shell thickness for a 0.30-pct. carbon steel cast in a single-taper, double-taper and constrained mould.

Fig. 10.   Model-predicted axial profiles of shell thickness and mid-face temperature for 0.13- and 0.30-pct. carbon steels cast in a single-taper mould.

versus-time profile, determined from mould temperature measurements, over this period then is virtually independent of shell thickness. Thus for any casting speed, the heat flux at a given distance below the meniscus for the 0.30-pct. carbon steel under consideration may be determined by computing the corresponding time t (= $\frac{d}{V}$) and obtaining the heat flux from Fig. 9.

For times greater than 7.8 s, the resistance of the shell exerts a significant influence on the overall heat extraction and must be included in the determination of a general relationship linking heat flux and time in the mould. Such relationships are established simply by incorporating the variation of the resistance of the shell with time in Eq.(1). For times greater than 7.8 s, the resistance of the shell is as follows:

7.8 < t < 11.0 s

$$R_s = \frac{0.905 \ t^{0.904}}{k_s} \tag{7}$$

t > 11.0 s

$$R_s \ \frac{1.46 \ t^{0.706}}{k_s} \tag{8}$$

The resistance of the gap for times greater than 7.8 s has been computed also and the results have been presented in Fig. 4. Thus the dependence of gap resistance on time in the mould can be characterized by the following

$$R_g = 0.1525 + 0.0005 \ Vt \tag{9}$$

Substituting Eqs.(7), (8) and (9) into Eq.(1) the variation of mould heat flux with time for the shell/gap resistance-controlled zone is obtained.

7.8 < t < 11.0 s

$$\dot{q} = \frac{T_s - T_w}{\frac{\delta_w}{k_w} + \frac{1}{h_c} + \frac{.905}{k_s} t^{0.904} + 0.1525 + .0005 \ Vt} \tag{10}$$

t > 11.0 s

$$\dot{q} = \frac{T_s - T_w}{\frac{\delta_w}{k_w} + \frac{1}{h_c} + \frac{1.46}{k_s} t^{0.706} + 0.1525 + .0005 \ Vt} \tag{11}$$

Equations (10) and (11) permit computation of heat-flux profiles in the lower part of the mould as a function of casting speed, which hitherto has been improperly accounted for in predicting tapers (Dippenaar, 1986). Obviously this procedure may be adopted for steel grades other than medium carbon; heat flux-time equations for a range of compositions including low carbon will be presented in a subsequent publication.

MOULD TAPER

Heat-extraction profiles obtained from mould temperature measurements with a single-taper mould have been employed with a mathematical model to predict the shrinkage of the solidifying shell and the desired taper (Dippenaar, 1986). The earlier study unambiguously demonstrated that a double-taper mould more closely matches the shrinkage of the solidifying shell than does a single-taper tube (Dippenaar, 1986). Moreover the tapers predicted with the model compared favourably with those currently being employed in industry. The heat-flux profiles that have been determined subsequently from measurement of mould temperature in a double-taper mould (Bommaraju, 1988) are different from those in a single-taper mould as is evident from Fig. 9. Hence it was considered worthwhile to recompute the shrinkage of the solidifying billet to assess whether the tapers currently employed are optimum. Taper is an important design variable and its influence on off-squareness and off-corner internal cracking has already been presented in previous publications (Bommaraju, 1984; Dippenaar, 1986; Samarasekera, 1982). In the next section, links also will be established between taper and the occurrence of transverse cracks and transverse depressions.

Figure 12 presents the midface temperature profiles computed for a 0.30-pct. carbon steel billet, 144-mm square, cast at a speed of 32.0 mm/s through single-taper, double-taper and constrained mould tubes. It is evident that there is a greater reduction in surface temperature for the single-taper and constrained mould billets in the upper portion of the mould as compared with the decrease for a double-taper mould billet. In the lower region, the billet surface temperature for the single-taper mould is closer to a condition of thermal steady state than the other two. These differences are reflected in the shrinkage profiles for each case shown in Fig. 13. These have been calculated according to the procedure outlined by Dippenaar (1986). From Fig. 13 it is seen that billets cast through a constrained or single-taper mould shrink considerably more than those cast through a double-taper tube over the upper zone which extends from the metal level at 175 mm down to a distance of 425 mm below the top of the mould. The average shrinkage associated with a single-taper or constrained mould billet over this region is 4.7%/m as compared with 4.10%/m for the double tapered mould billet. Below 425 mm, the average shrinkage of single-taper, double-taper and constrained mould billets are 0.92%/m, 1.76%/m and 1.76%/m. The effective taper that should be imparted to the mould to accommodate the shrinkage of the billet can be computed from these values by allowing for mould distortion as described by Dippenaar (1986). In the upper region, the mould bulges outward and, if untapered, acquires a negative taper from the meniscus down to the position of the maximum bulge and, below it, a positive taper. The actual magnitude of the distortion is strongly influenced by mould wall thickness, type of mould tube support, cooling water velocity, metal level, casting speed and steel grade (Samarasekera, 1982). Hence taper requirements will vary from one mould system to the next for a given steel grade. It is also clear from the billet surface temperature profiles that a continuously varying taper would be more effective than a double taper in compensating for the shrinkage of the billet.

To assess the effect of casting speed on billet shrinkage and taper requirements, computations have been made for a 0.30-pct. carbon 203-mm square billet cast at 20.3 mm/s and the results compared with the shrinkage of a 144-mm square billet cast at 32.0 mm/s. The heat-flux profile for the lower casting speed has been computed according to the procedure described in the previous section; the results are shown in Fig. 14. For the lower speed the average heat-extraction rate in the shell/gap resistance-controlled zone is

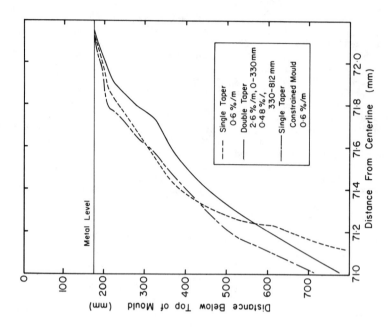

Fig. 13. Model-predicted axial profiles of billet dimensions for a 0.30-pct. carbon steel cast in a single-taper, double-taper and constrained mould.

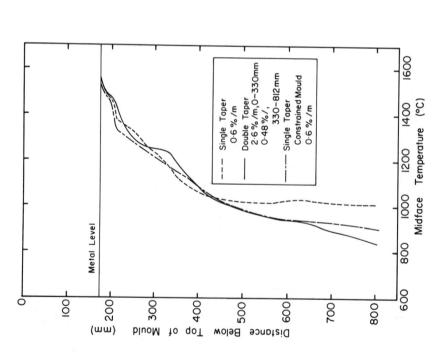

Fig. 12. Model-predicted axial profiles of mid-face temperature for a 0.30-pct. carbon steel cast in a single-taper, double-taper and constrained mould.

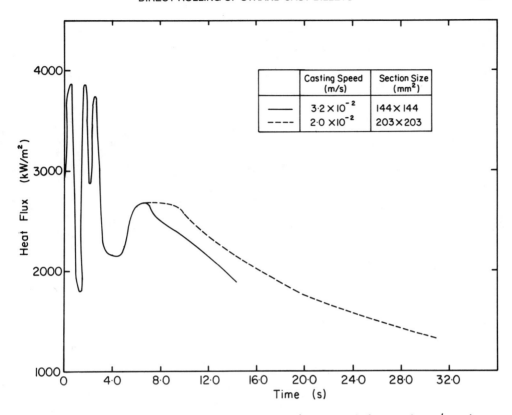

	Casting Speed (m/s)	Section Size (mm^2)
——	$3{\cdot}2 \times 10^{-2}$	144×144
----	$2{\cdot}0 \times 10^{-2}$	203×203

Fig. 14.   Change of mould heat flux with time (distance below meniscus/casting speed) for casting speeds of 20.3 mm/s and 32 mm/s for a 0.30-pct. carbon steel.

lower than that at the higher speed owing to the increased resistance associated with the thicker shell in the former case. This is illustrated in Fig. 15 which shows the shell thickness and surface temperature profiles for the two cases under consideration.   As expected, the shell thickness is considerably greater for the slower casting speed and the difference increases with distance below the metal level. At the lower casting speed the surface temperature is virtually constant over the region 500 ~ 800 mm below the top of the mould, a direct result of the lower heat-extraction rates.

The temperature distribution in the growing shell has been employed to calculate the shrinkage and associated billet surface profile for the two cases; the results are shown in Fig. 16. In both cases the magnitude of the shrinkage is higher, in the upper region of the mould than in the lower region as expected. However, there are some significant differences in the magnitude of the shrinkage which is of paramount importance in designing taper. For the casting speed of 20.3 mm/s the billet dimensions decrease linearly with distance below the metal level in the upper zone and the shrinkage is ~ 5.53%/m over this region which extends to a depth of 370 mm below the top of the mould.   For the higher casting speeds the billet dimensions decrease non-linearly with distance below the metal level in the upper zone which

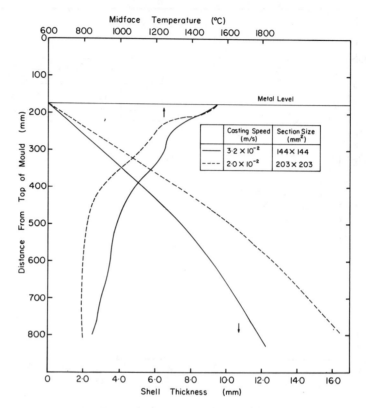

Fig. 15. Model-predicted axial profiles of shell thickness and mid-face temperature with casting speeds of 20.3 mm/s and 32 mm/s for a 0.30-pct. carbon steel.

extends from the metal level to 425 mm below the top of the mould, and averages 4.01%/m. Below the first zone the taper requirements for the low casting speed case decrease continuously with distance below the breakpoint. Over the 500 ~ 800 mm region, for instance, the shrinkage is approximately 0.58%/m so that the actual taper required would be only approximately 0.18%/m after the thermally generated positive taper of ~ 0.4%/m acquired by the mould is subtracted. In contrast, for the higher casting speeds the average shrinkage below 425 mm is approximately 1.76%/m and the required taper would be approximately 1.36%/m. The predicted reduction in second taper with the low casting speeds is a direct result of the reduced heat extraction in the lower part of the mould stemming from the increased resistance to heat flow that accompanies the thicker shell. This is contrary to current practice in industry in which both the first and second tapers of a double tapered mould are increased for larger sections which are generally cast at lower speeds. This finding has implications with respect to the formation of transverse depressions and cracks which will be discussed in the ensuing section on billet quality.

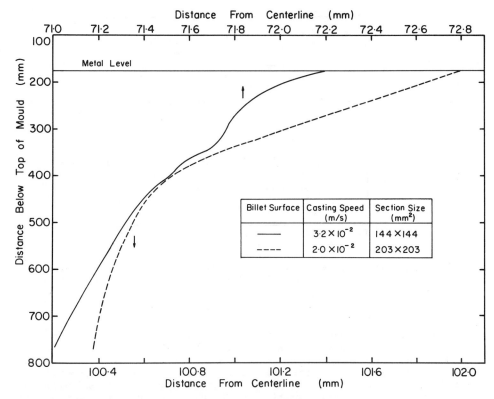

Fig. 16. Model-predicted axial profiles of billet dimensions for a 0.30 pct.
carbon steel cast at 20.3 mm/s and 32 mm/s.

IMPLICATIONS FOR BILLET QUALITY

Earlier studies have elucidated the influence of heat extraction, thermomechanical behaviour of the mould and mould/shell interaction on the generation of off-squareness and off-corner, internal cracks in billets (Bommaraju, 1984; Brimacombe, 1984; Brimacombe, 1986; Dippenaar, 1986, Samarasekera, 1982; Samarasekera, 1984). It has been shown that mechanical interaction between the mould and newly solidified shell at the meniscus during the negative-strip period of the mould oscillation cycle can lead to off-squareness if oscillation marks that form at this time are nonuniform around the periphery of the billet. The shell/mould gap and heat transfer then vary in the transverse plane so that shell growth is non-uniform. Under these conditions, the billet emerging from the mould can be pulled off-square by the sprays due to differential cooling. Increasing the taper close to the meniscus reduces the shell/mould interaction and improves the uniformity of heat extraction to the extent that off-squareness is reduced. Similarly maintaining the correct taper down the mould and preventing uncontrolled mould distortion (high cooling water velocity, excellent water quality) enhances uniform heat extraction which again improves off-squareness problems. Recent evidence of the influence of mould taper on off-squareness is shown in Fig. 17, taken from a study by Lorento (1988).

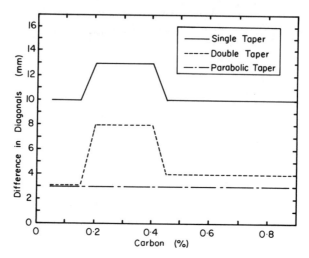

Fig. 17. Influence of mould taper on off-squareness over a range of carbon contents (from Lorento, 1988).

Taper also affects the severity of off-corner internal cracks which form due to bulging of the steel shell in the lower region of the mould. In combination with deep oscillation marks and a thin shell at off-corner sites, the bulging causes a hinging action locally at these locations and the generation of tensile stresses in the zone of low ductility adjacent to the solidification front. The result is crack formation in the off-corner regions. Consequently prevention of excessive shell/mould gaps and achievement of uniform heat extraction by proper taper and minimization of mould distortion are as effective potentially in reducing off-corner internal cracks as they are in remedying off-squareness.

Also of significance to billet quality is the role of mould heat extraction and taper in the generation of transverse depressions and cracks. These quality problems have cropped up with increasing frequency as refinements to mould taper continue to be made. The mechanism by which these defects are generated in the mould is illustrated schematically in Fig. 18. Shown is a longitudinal section through the shell at an instant when it is sticking to, or binding in, the mould tube. Under these conditions of high local mould/shell friction, the shell is subjected to a high axial tensile stress due to the mechanical pulling of the withdrawal system. Depending on the magnitude of the stress, the shell can begin to flow plastically and form a neck, much as in a standard laboratory tensile test, because at temperatures typical of those in the mould, between about 1150 and 1430°C, steel has a high ductility (Thomas, 1986). The neck is manifested as a depression on the billet surface as shown in Fig. 18. Close to the solidification front, within about 50°C of the solidus temperature, however, the steel has virtually zero ductility so that, under the influence of tensile strains, a transverse crack forms, Fig. 18. Depending on the shell thickness at the time of crack formation, as well as the extent of necking, the crack may penetrate to the surface. Thus it is common for transverse cracks to be found at the bottom of transverse depressions. From Fig. 18 it is apparent that the depth of a crack that has not propagated to the surface gives a rough measure of the shell thickness at the time the crack formed.

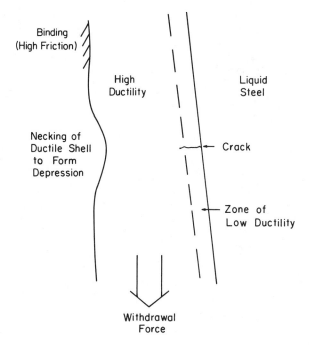

Fig. 18. Schematic diagram showing the formation of
transverse depressions and cracks in billets
due to binding in the mould.

Binding or sticking of the billet in the mould can have one or more origins:

[1] mould shape – improper taper (sometimes related to casting speed and metal
     level) and permanent mould distortion;
[2] lubrication – type, flow rate and distribution;
[3] alignment between mould and sub-mould support system;
[4] mould oscillation – wobble, vibration.

More will be stated on the role of lubrication in future papers but for the
present, our comments will be confined to the question of mould taper.

Clearly, excessive taper will result in binding between the billet and mould
tube with the generation of transverse defects. From the findings of the
present study, the problem of binding will increase as casting speed is
reduced. This is easily explained by the fact that as the dwell time increases
with decreasing speed, the influence of the shell conduction resistance grows,
the heat flux declines (Eqs.(10) and (11) and Fig. 14) and the cooling
rate/shrinkage falls off, eg. Fig. 15. Thus for a given taper, the reduced
shrinkage may result in binding. If this is the case, the most effective
immediate remedy is to increase casting speed and/or drop the metal level to
reduce the dwell time of the steel in the mould.

As was presented earlier, at a casting speed of 20.3 mm/s, the taper in the
lower region of the mould needs only to be about 0.18%/m. This value is
considerably smaller than tapers presently being applied for such casting

speeds; indeed current practice is to increase the lower taper of double-taper moulds as casting speed is decreased. Clearly a reduction of taper based on measured heat fluxes and calculated shell shrinkage should be considered and is the subject of a current study.

On a final note, binding will tend to be worse when casting low-carbon billets (0.1-0.14 pct. C) because less heat is transferred to the mould, Fig. 6; consequently, cooling and shrinkage of the shell are less than in the case of a medium-carbon billet, Fig. 10. However, steels falling in the range of 0.17 to 0.24-pct. carbon also may exhibit a higher frequency of transverse defects owing to a diminished ductility at elevated temperatures (Brimacombe, 1977).

SUMMARY

The heat-extraction capability of continuous-casting billet moulds has been examined by comparing heat-flux profiles derived from mould temperature measurements on three mould systems: a single-taper mould, a double-taper mould, and a specially designed constrained mould in which, unlike conventional mould systems, transverse expansion is restrained. The salient aspects of the study are summarized below.

[1] The mould can be divided into two zones: an upper region of gap-resistance dominance in which heat extraction can be influenced by factors altering the gap width like taper and mould distortion and a lower region of gap and shell resistance dominance in which heat extraction can be influenced by factors like casting speed (alters the shell thickness) and taper (changes gap). For a 0.30-pct. carbon steel the transition between the zones was found to occur at a shell thickness of 6 mm which corresponds to a time of 7.8 s (time = distance below the meniscus/casting speed).

[2] On comparing the heat-extraction profiles for the single-taper, double-taper and constrained moulds, it is clear that mould distortion has a significant influence on heat transfer in the upper zone. The single- and double-taper mould heat-extraction profiles exhibit 2 ~ 3 maxima and minima which are absent in the constrained mould heat-extraction profile and can be associated with mould distortion.

[3] Shell growth, mid-face temperature and shrinkage profiles have been computed for the three mould systems with a mathematical heat-flow model. It is evident that in the upper zone, the surface temperature decreases rapidly and is accompanied by substantial shrinkage whilst in the lower zone, where the shell resistance is significant, the billet approaches a thermal steady state in which the shrinkage is significantly less. Hence a continuously decreasing taper would be appropriate for compensation of shrinkage.

[4] To permit determination of the influence of casting speed on mould heat transfer, a relationship linking heat transfer and time in the mould, which contains an expression for shell resistance, gap resistance, mould wall and mould cooling water resistances has been formulated for a 0.30-pct. carbon steel.

[5] The billet surface, shell growth and shrinkage profiles have been calculated for the 0.30-pct. carbon steel billet having a section size of 203 mm square cast at 20 mm/s and for a section size of 144 mm square cast at 32 mm/s. In the upper zone, a larger taper is required for the large section with the low casting speed whilst in the lower zone the taper requirements for this case are much less than for the smaller section cast at the higher speed. This stems directly from the reduced heat extraction lower in the mould with the low casting speeds owing to the thicker shell.

[6] It has been postulated that transverse depressions and cracks form in the mould due to mechanical pulling of the withdrawal system on the shell when binding or sticking occurs. This imposes a longitudinal tensile stress on the shell such that a transverse crack can form close to the solidification front where the ductility is extremely low. At the surface the material flows plastically and forms a depression much like the necking phenomenon in a laboratory tensile test.

[7] The binding or sticking could be particularly severe for low-carbon steels (0.1 ~ 0.14 pct. carbon) because less heat is transferred to the mould due to wrinkling of the shell. Similar problems could occur when casting large sections at low speeds with improper lower taper. The tapers currently employed for the lower region of the mould are too steep relative to the findings of this study.

ACKNOWLEDGEMENTS

The authors wish to express their appreciation to Dr. R. Bommaraju and Mr. N. Walker for their efforts. The financial support of NSERC is gratefully acknowledged. The in-plant measurements, which are a corner-stone of this study, could not have been accomplished without the co-operation and support of numerous steel companies.

NOMENCLATURE

d	Distance below the metal level (mm)
h_c	Heat-transfer coefficient at mould/cooling water interface (kW/m^2°C)
k_s	Thermal conductivity of steel (kW/m°C)
k_w	Thermal conductivity of mould wall (kW/m°C)
$\dot{q}$	Heat flux from steel to mould (kW/m^2)
R_g	Resistance of gap to heat flow (m^2°C/kW)
R_s	Resistance of shell to heat flow (m^2°C/kW)
R_T	Total resistance to heat flow (m^2°C/kW)
t	Time in the mould (s)
T_s	Solidus temperature of steel (°C)
T_w	Cooling water temperature (°C)
V	Casting speed (mm/s)
δ_w	Mould wall thickness (m)
δ_s	Shell thickness (m)

REFERENCES

Berryman, R., Samarasekera, I.V. and Brimacombe, J.K. (1988). Cooling Water
Flow in Continuous-Casting Billet Moulds. ISS Trans., Vol. 9, No. 3.

Bommaraju, R., Samarasekera, I.V. and Brimacombe J.K. (1984). Mould Behaviour
and Solidification in the Continuous Casting of Steel Billets.
III. Structure, Solidification Bands, Crack Formation and Off-Squareness",
ISS Trans., 5, 95-105.

Bommaraju, R. (1988). Mould Behaviour, Heat Transfer and Quality of Billets
Cast with In-Mould Electromagnetic Stirring. Ph.D. Thesis, The University of
British Columbia.

Brimacombe, J.K. (1976). Design of Continuous Casting Machines Based on a
Heat-Flow Analysis: State-of-the-Art Review.   Can. Met. Quart., 15,
163-175.

Brimacombe, J.K. and Sorimachi K. (1977).   Crack Formation in the Continuous
Casting of Steel. Metall. Trans. B, 8B, 489-505.

Brimacombe, J.K., Samarasekera, I.V., Walker, N., Bommaraju, R., Bakshi, I.,
Hawbolt, E.B. and Weinberg, F. (1984). Mould Behaviour and Solidification in
the Continuous Casting of Steel Billets. I. Industrial Trials. ISS Trans.,
5, 71-77.

Brimacombe, J.K., Samarasekera, I.V. and Bommaraju, R. (1986). Optimum Design
and Operation of Moulds for the Continuous Casting of Steel Billets. Fifth
Intl. Iron and Steel Congress, Washington, D.C., Steelmaking Proc., 69,
409-423.

Dippenaar, R.J, Samarasekera, I.V. and Brimacombe, J.K. (1986). Mould Taper in
Continuous Casting Machines. ISS Trans., 7, 31-43.

Grill, A. and Brimacombe, J.K. (1976). Influence of Carbon Content on Rate of
Heat Extraction in the Mould of a Continuous Casting Machine. Ironmaking and
Steelmaking, 2, 76-79.

Hurtuk, D.J. and Tzavaras, A.A. (1982). Solidification and Casting of Metals.
J. Metals, Vol. 34, No. 2, 40.

Kreith, F. and Black, W.Z. (1980). Basic Heat Transfer, Harper and Row Publ.,
New York, 52-54.

Lorento, D.P. (1988). Development in Mould Technology. Paper presented at
Globe-trotters meeting.

Pugh, R.W., Staveley, P.R., Samarasekera, I.V. and Brimacombe, J.K. (1988).
Mould and Oscillator Changes for Improved Billet Quality and Caster
Performance. Proc. of Intl. Symposium of Direct Rolling and Direct Charging
of Steel Billets, CIM Conference of Metallurgists, Montreal.

Samarasekera, I.V. and Brimacombe, J.K. (1979).   The Thermal Field in
Continuous Casting Billet Moulds. Can. Met. Quart., 18, 251-266.

Samarasekera, I.V. and Brimacombe, J.K. (1982).   Thermal and Mechanical
Behaviour of Continuous Casting Billet Moulds. Ironmaking and Steelmaking,
9, 1-15.

Samarasekera, I.V., Brimacombe, J.K., Bommaraju, R. (1984). Mould Behaviour
and Solidification in the Continuous Casting of Steel Billets. II. Mould
Heat Extraction, Mould-Shell Interaction and Oscillation-Mark Formation. ISS
Trans., 5, 79-94.

Singh, S.N. and Blazek, K.E. (1976). Heat-Transfer Profiles in a Continuous-
Casting Mould as a Function of Various Casting Parameters. Open Hearth Proc.
AIME, 59, 264-283.

Thomas, B.G., Brimacombe J.K. and Samarasekera, I.V. (1986). The Formation of
Panel Cracks in Steel Ingots: A State-of-the-Art Review. I. Hot Ductility
of Steel. ISS Trans., 7, 7-18.

Wolf, M. (1980). Investigation into the Relationship Between Heat Flux and
Shell Growth in Continuous-Casting Moulds. Trans. ISIJ, 20, 710-717.

SESSION 2

TEMPERATURE EQUALIZATION METHODS AND EQUIPMENT

Chairpersons: **B. Bowman (Slater)**
                         G.E. Ruddle (CANMET-MTL)

INDUCTION HEATING OF BILLETS IN DIRECT ROLLING

G. Bendzsak, and J. Cox
Hatch Associates, Toronto, Ontario

ABSTRACT

This paper describes induction heating of steel billets in a direct rolling
application.   It is shown that both the design and operation of the process
requires a thermal analysis of the billet from its liquid state to its entry into
the rolling mill. The paper outlines the interactions between process units and
relates these to heating requirements. Various design and operating considerations
for a practical direct rolling line are discussed, and the results from a
comprehensive mathematical model are presented to illustrate the role of
induction heating in direct rolling.

KEYWORDS

Induction heating; temperature compensation; direct rolling; system operation;
control requirements; power consumption;

INTRODUCTION

In conventional steelmaking practice, after a billet leaves the caster it cools
to the ambient temperature. Prior to rolling, it is placed into a re-heat furnace
where it is brought up to temperature. This energy intensive step is dictated by a
number of considerations which include the need for inspection, conditioning,
scheduling and coordination of unit operations.

During recent years, all aspects of steelmaking have seen changes that gradually
increased the quality of cast steel products. Consequenly, the need for billet
inspection and conditioning has decreased. Computer systems invaded all aspects of
steelmaking to achieve an increasingly better coordination of the operations and
its process units. As a result, today it is possible to consider direct rolling as
a viable technology awaiting a more frequent implementation. The principle reward
for its realization is a substantial reduction in the costs of energy associated
with re-heating.

The average temperature of a billet, on its exit from a caster, is usually below that required for rolling. As a consequence of rapid cooling within the caster, the temperature profiles manifest steep gradients in both cross section and logitudinal direction. These characteristics, combined with a need to accomodate short delays, dictate the installation of a heating station between the caster and the mill. This heating facility must respond both rapidly and automatically to changing plant operating conditions. Induction heating liberates power within the body of the billet to allow the realization of rapid heating rates. Furthermore, the power can be instantaneously adjusted to any desired level. These features render induction heating well suited for its incorporation into a flexible direct rolling system.

A DIRECT ROLLING INSTALLATION

The design and analysis of a heating system for direct rolling must take into account the layout of a particular facility. Typical system components and a possible arrangement are shown in Figure 1. This layout shall be used as a reference throughout the discussions of the paper. The system is assumed to consist of a four strand billet caster, a delivery system to the induction heaters, and a discharge unit located between the heaters and the rolling mill.

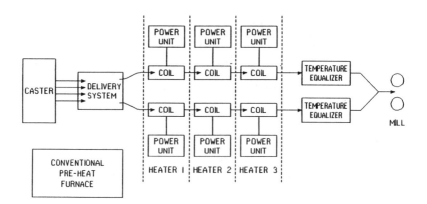

Figure 1. Direct Rolling System Schematic

For a baseline case, the system has the following characteristics:

   a.) The caster can produce 150 mm by 150 mm billets at 2.0 m/min for an overall production capacity of 80 tons/hr. The nominal billet length is 12 m.

   b.) The transit time between the straighteners in the caster and cut-off torch is 3 minutes for a 150 mm by 150 mm billet.

c.) The delivery system between the caster and the heating station may have covers installed to maintain billet temperatures during a 6 minute transit time.

d.) The billets are delivered to a covered area just ahead of the heating station, where they spend negligible time. This area can be used as a temporary depot only in case of brief mill delays.

e.) There are two parallel induction heating lines, consisting of three heating stations each. The average billet temperature can be raised from 1000 $^{\circ}$C to 1200 $^{\circ}$C at a speed of 6.0 m/min. This means a combined heating capacity of 120 tons/hr, which is well above the production capability of the caster to allow for extra heating capacity after the event of a delay.

f.) On exit from the heaters, the billets enter a covered transit zone for 2 minutes to allow for better temperature equalization.

A typical section of an induction heating system consist of a 1.0 m long, single layer coil. The billets pass through the center of the coils and are supported by rollers placed in between the heater sections. While the design of the coils can be optimized for one billet size, they will operate over the whole range of product sizes. The performance is expected to be excellent over the cross sectional areas (100 mm by 100 mm to 150 mm by 150 mm) typically produced by a billet caster.

The depth to which the induced currents penetrate a billet can be altered by a change in the frequency of applied power. This allows a direct control over the cross sectional area where the power is liberated. By an adjustment to the frequency of the source it is possible to control the shape of the resultant temperature profiles. This aspect of induction heating makes the power supply an important installation feature .

During regular operation, the induction heating system is expected to encounter many unexpected situations. Accordingly, it is desirable to implement a most flexible system suitable for the realization of a control strategy. This can be achieved when each coil is energized by a separate solid state power supply controlled by a computer. The nominal frequency of the power source is 1000 Hz.

For the purposes of analysis, it is assumed that under normal operating conditions, all cast products will be heated directly prior to delivery to the rolling mill. Such an operation in practice could not be realized for a number of reasons. These include the removal of defective billets for conditioning, excessive heat loss during long delays, and startup. Consequently, there would be always a requirement to heat steel from cold. An induction heating system is designed to be charged with hot steel and would not have sufficient rating to handle the high power demand of cold, magnetic billets.

The rating of an additional heating system would depend on expected operating characteristics of the plant. Based on the experience of a number of steel plants that attempted direct rolling in the past, over 80% of total production can be rolled directly on a consistent basis. Thus, a reasonable estimate for the size of the additional heating unit would be in the order of 20% of production capacity. The billets need to be raised to an average temperature in the order of 1000 $^{\circ}$C, as the final adjustments can be done by the induction heating system.

SYSTEM MODELS AND ASSUMPTIONS

The heating system must deliver billets to the rolling mill that have near uniform temperature profiles. It is the role of the induction heater to raise the billet bulk temperature as well as to compensate for any non-uniformity in those profiles. During operation only surface temperatures can be measured directly. The status of internal temperature profiles can be estimated only by references to measurements made under laboratory conditions or to results of calculations based on analytical process models. While laboratory measurements are crucial to success, these are both expensive and difficult to obtain for a wide range of operating conditions. The physical laws required for a mathematical description of direct rolling are well understood. Based on the laws of heat transfer and electromagnetics, detailed models can be constructed to provide both accurate and rapid assessment of billet temperature profiles under all operating conditions.

The physical processes that should be included in the computational models are

1.) freezing and cooling of the steel in the caster
2.) cooling of a billet during transit between the caster and the induction heaters
3.) electromagnetics of the induction heaters
4.) thermal anaysis of billets during transfer to the rolling mill

The temperature distribution within the billet can be calculated for any externally defined condition from the conservation of heat. This is given by the following vector differential equation (Szekely, 1971):

$$\rho \; C_p \frac{\partial T}{\partial t} = \nabla \cdot \left(k \; \nabla T \right) + P \qquad \text{1.)}$$

where: T = temperature
           t = time
           ρ = density of steel
           P = electrical power density
           ∂ = differential operator
           k = thermal conductivity
           C_p = heat capacity

In order to specify the thermal problem completely it is necessary to include the effect of heat loss through the billet surfaces under all operating conditions. The two heat flux mechanisms that must be considered are radiation and convection. When the analysis involves the caster, additional care must be taken to regarding the following:

a.) large increase in the value of heat capacity for steel during transition from the solidus to liquidus temperatures.
b.) large heat flux through the mould surfaces
c.) effective convective heat transfer coefficients for spray zones
d.) emmissivity of the billet surfaces
e.) heat extraction due rollers

For results presented in this paper, the inside surface temperature of any cover in a transit zone is equal to the average temperature of the billet, slightly adjusted for heat losses. Since the covers are assumed to be made of good insulating material, this loss component has been taken to be small. It was found, however, that the the results for billet temperature profiles are insensitive to variations in this quantity.

The inside surface temperature of an induction heater is difficult to calculate, hence it was assumed that it was maintained 75 °C below that of the average billet temperature on exit from the coil. Once again, the billet temperature profiles are not very sensitive to variations in this quantity.

The thermal aspects of the process represent a part of the problem. The electrical behaviour of the induction heaters can be described by Maxwell's field equations under standard low frequency approximations. These equations are (Smythe 1968):

$$\nabla \times \overline{H} = \overline{J} \qquad\qquad 2.)$$

$$\nabla \times \overline{E} = -\partial \overline{B}/\partial t \qquad\qquad 3.)$$

$$\nabla \cdot \overline{E} = \rho / \varepsilon \qquad\qquad 4.)$$

$$\nabla \cdot \overline{B} = 0 \qquad\qquad 5.)$$

where: H = magnetic field intensity
B = magnetic flux density
E = electric field intensity
J = current density
= magnetic permeability
= electrical conductivity
= electric permittivity

If the magnetic field density B is defined in terms of a vector potential as $B = \nabla \times A$, the overall relationship becomes:

$$\nabla \times \frac{1}{\mu} (\nabla \times \overline{A}) = -\sigma \nabla \emptyset - \sigma \, \partial \overline{A}/\partial t \qquad 6.)$$

When the driving force is sinusoidal at a frequency w, equation 6.) reduces to:

$$\nabla \times \frac{1}{\mu} (\nabla \times \overline{A}) = -\overline{J} - j \omega \sigma \overline{A} \qquad\qquad 7.)$$

The solution of the above equations for both heat conduction and electromagnetics is not possible by analytical methods because the boundary conditions and material properties are all temperature dependent. Thus, they must be transformed into finite difference or finite element forms whose algebraic equations can be solved by numerical methods(Patankar). The results presented in the following sections will show that the combination of the equations of electromagnetics and conservation of thermal energy gives realistic billet temperature profiles for an engineering analysis of the direct rolling process.

RESULTS OF THE ANALYSIS

A billet with a cross sectional area of 150 mm by 150 mm was chosen to illustrate the major features of a direct rolling system using induction heaters for temperature compensation. This billet size is a representative of the larger pieces of steel to be passed through the heater, and can be used as a guide for its maximum rating.

It should be stated at the outset that the major driving force for direct rolling is the energy savings in re-heating. It is important that billets be cast with high average temperatures. In the chosen example, the rate of heat extraction in the mould and the spray zones were adjusted such that the approximate average temperature at the caster exit point is 1125 °C. Through proper adjustment of the cooling system the cross sectional temperature profiles are symmetrical about the center lines. As can be seen in Figure 2. the billets have steep temperature gradients and attention must be paid to the minimum surface and corner temperatures to avoid cracking during the straightening process. As the billet leaves the caster, heat transfer rates through the surfaces become lower and the energy leaving the billet central zone re-heats the outer regions. This temperature rebound can induce stresses that may cause quality problems. Billet inspection and conditioning represent one of the main problems for direct rolling and should be minimized as much as possible. Temperature control of billets, in the caster and immediately after it, represents one of the key considerations in the implementation of a successful direct rolling operation.

Covers for billets during transit between various process units are believed to be the cause of operating problems. Thus in the chosen example, the billets were transferred to the heating zone without these simple energy saving devices. The total transit time between caster exit point and arrival at the heating stations was taken to be 9 minutes. During this time the billet lost heat through radiation in a uniform manner. Figure 3. shows that at the point of entry to the heater, the average temperature has dropped to 992 °C, and the temperature profiles are almost uniform.

Before a billet can enter the first coil of the induction heater its straightness in the longitudinal direction must be determined. Design of efficient coils dictates the distance between it and the load to be as small as possible, reducing a heating system's tolerance to billet bending. The damage caused by the feeding of a crooked billet into the heater would put the whole operation into serious jeopardy. Thus, installation of billet straighteners may be required before the heating station.

The case under consideration also points towards a major problem introduced by small delays in the mill. Since the billet lost energy in a thermally unprotected environment, the temperatures begin to approach levels that are borderline cases for re-heating. In the present case, should a delay occur in excess of 1 minute, the temperatures would drop too low for re-heating at 120 tons/hr. The rating of the heater was set above that of the caster just to take care of these situations. In this hypothetical, but not unlikely, case the entire direct rolling facility would be in danger of never being able to catch up with the caster. Thus the prevention of significant heat loss by means of simple covers presents a buffer to allow for delays of longer durations. In addition, this example also underscores the importance of the ability to cast billets with average temperatures above 1125 °C.

The billet is heated by three coils operating at a frequency of 1000 Hz. Since the surface temperatures of the billet are low, it is safe for each of the first two coils to supply 820 KW to the billet. Taking losses into account, this results in average heating rates in excess of 7 °C per second. Figures 4. and 5. show typical cross sectional temperature profiles at the exit of the first or the second coil. At the chosen power frequency, the induced current effectively liberates power in the first 15 mm from the surface. This zone reaches approximately 1275 °C after the second coil. It becomes evident that heating at these rates would result in the possible melting of the outside steel layers. Thus in the third coil the power is reduced to supply 700 KW into the billet. The average exit temperature from the last coil is 1195 ^{o}C and the maximum at the surface is 1325 ^{o}C. It is interesting to note that the central temperature has increased only by 25 ^{o}C while the billet was in the induction heater. The steep temperature gradients will diffuse thermal energy into the body of the billet and a period of equalization begins.

The speed of the billet was 6 m/min, allowing a 2 minutes passage through the heater. During this time the control system must adjust power continually to take into account longitudinal temperature variations. When the billet is through the heater, the center line should be allowed to come up to temperature. The best way to achieve this is to place the billet into an enclosure for 2 minutes. At the end of this period, the billet temperatures become completely homogeneous as shown in Figure 7. The average temperature is approximately 1170 ^{o}C, while the maximum deviations are less than 10 ^{o}C from center to corner and less than 5 ^{o}C from mid-face to corner. These differentials are well within the required tolerances of a rolling mill.

The time history of the heating and equalization periods for the base case is shown in Figure 8. The importance of a covered transit is shown in Figure 9. The starting point for the steel is identical to that of the previous case, but the billets are transported to the heater in a covered environment. All temperatures are significantly higher than in the previous case. This by itself would allow delays in the rolling mill up to 6 minutes in duration. There is an additional benefit in that the power required to reach the same end point is approximately 80% of the base case. This allows for a further extension of the delay period. Thus the limiting factor is no longer the induction heater but the size of the storage area in front of it.

The importance of the chamber after the heaters is shown in Figure 10. In this case the process is identical to that shown in Figure 9. with the removal of the temperature equalization chamber. The billet is allowed to radiate to the ambient, thereby losing heat rapidly through its surface. By the time the billet reaches the rolling mill, the temperature differences are in the order of 50 ^{o}C. This is too large a value to be tolerated by the mill.

CONCLUSIONS

The results of the analysis in this paper show that the performance evaluation of an induction heater in direct rolling system for billets cannot be conducted in isolation. The success of the line depends on the interplay between all major system components from the caster to the rolling mill. The major conclusions of the paper are that :

a.) an induction heating system can be designed to meet the heating requirements of direct rolling

b.) the caster should produce billets having as high an average temperature as possible.

c.) the billets should be shielded against radiation losses by covers installed over all delivery systems between major equipment.

Induction heating provides a fast and flexible system for the adjustment of billet temperatures in a direct rolling application, provided its operation is coordinated with both up and downstream process units.

REFERENCES

Szekely, J.(1971). Rate Phenomna in Process Metallurgy. Wiley,
    New York.
Smythe, W.R.(1968). Static and Dynamic Electricity. McGraw-Hill,
    New York.
Patankar, S.V. (1980). Numerical Heat Transfer and Fluid Flow. Hemisphere
    Publishing Corporation, New York.

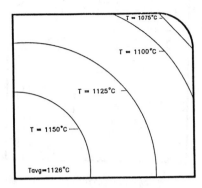

Figure 2.   Temperature profiles at exit from caster

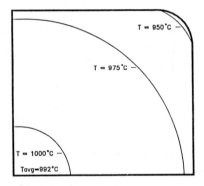

Figure 3.   Temperature profiles at entrance to heater

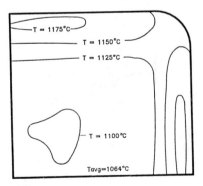

Figure 4.   Temperature profiles after first coil

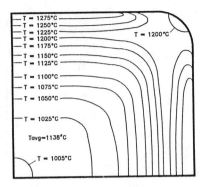

Figure 5. Temperature profiles after second coil

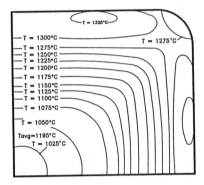

Figure 6. Temperature profiles after third coil

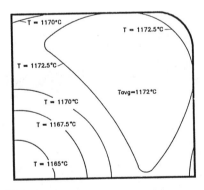

Figure 7. Temperature profiles at mill entry

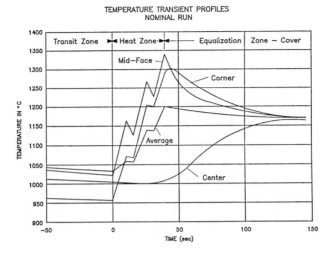

Figure 8.   Temperature transients for base case

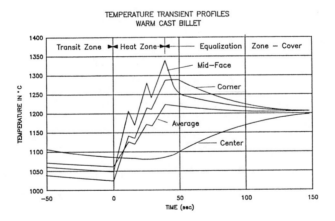

Figure 9.   Temperature transients with cover to heater

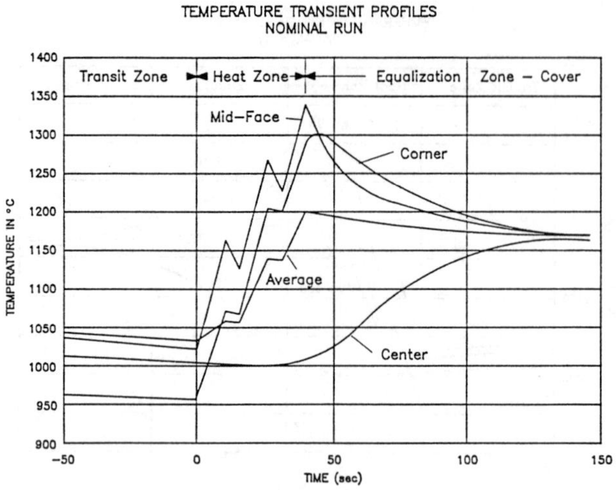

TEMPERATURE TRANSIENT PROFILES
NOMINAL RUN

Figure 10.   Temperature transients - no cover to mill

INDUCTIVE INTERMEDIATE REHEATING IN ROLLING MILLS

Ulrich Haldenwang

Asea Brown Boveri AG, Dortmund

Abstract

Optimum temperature control during hot rolling is one of the prerequisites for increasing economic efficiency and product quality in a rolling. Induction heating offers special benefits for temperature control in rolling mills, due to the high power concentration achieved by generating heat directly in the workpiece.

Introduction

Optimum temperature control is one of the basic conditions for increasing profitability and product quality in a rolling mill.

Increasing product quality means:

- reducing surface decarburisation and scale formation,
- obviating coarse grain formation and intercrystalline corrosion,
- reducing surface defects,
- restricting the range of tolerances for cross sections.

Increasing profitability means:

- rapidly adjusting the billet temperature for rolling programs involving frequently changing qualities and small lot sizes,
- saving energy by utilizing residual heat from previous processes,
- selective reheating of the stock during the rolling processes.

In this instance induction heating affords special possibilities and advantages for temperature control in the rolling mill because of the high concentration of power by generating the heat directly in the work piece.

Setup of ABB Induction Heating Systems

An induction heating plant consists essentially of the following functional groups (Fig. 1):

1. Mains supply into the converter transformer. Power is supplied directly from the existing medium voltage system of the plant.
2. Static frequency converter to supply the heater with the required medium frequency power.

3. Material feeder for the heater.
4. Heating facility with capacitor bank, control section and induction heating coil.
5. Equipment for material transfer to the following shaping facility.
6. Recooling installation for dissipating the heat losses from the converter, the capacitor bank and induction heating coils.

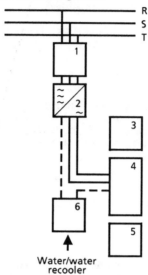

Fig. 1 Block diagram of a complete induction heating installation with the associated auxiliary equipment.

Basic Principals of Induction Heating

Heating in the induction heating coil is effected by alternating currents of a medium frequency induced in the material itself. The water cooled induction coil, which is designed as a hollow copper section coil, remains nearly cold. To minimize radiation losses of the heated material, and for electrical insulation, the inductor is lined with refractory material. In the inductive heating process the heating coil acts like the primary winding and the material to be heated like the single-turn short-circuited secondary winding of a transformer (Fig. 2).

Because of what is known as skin effect, a heavy heating current flows in the surface layer of the material to be heated. The thickness of this heated layer generated by the heating current is a function of the frequency used in the process. The higher the frequency, the smaller is the depth of current penetration or the heated surface layer.

As most of the heat is generated in the surface layer of the material, the core of the stock is heated by heat conduction. The selection of the optimum frequency is a compromise between optimal electrical and optimal thermal efficiency (Fig. 3). The smaller the current penetration depth, i.e. the thinner the heated layer, the higher is the electric efficiency.

In practical operation, the most favourable power transfer is obtained with a billet diameter of about 3.5 times the current penetration depth. The economical range is between 3 and 7 times the current penetration depth.

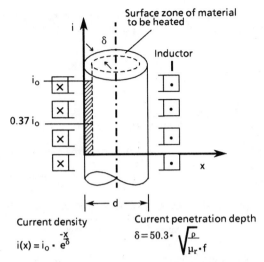

Current density

$$i(x) = i_o \cdot e^{\frac{-x}{\delta}}$$

Current penetration depth

$$\delta = 50.3 \cdot \sqrt{\frac{\rho}{\mu_r \cdot f}}$$

Fig.2 Current density and penetration depth in inductive heating of a metal cylinder.

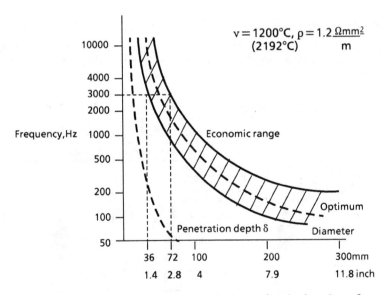

Fig.3 Frequency as a function of diameter for the heating of steel above Curie point.

Possible Applications of Induction Heating

Application of induction heating for optimum temperature control is possible in the production phases after casting / before rolling as well as in the rolling process (Fig.4).

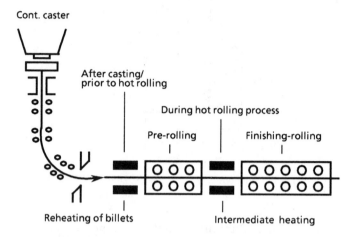

Fig. 4 Induction heating application in rolling mills.

Temperature Control in the Operation Phase after Casting / before Rolling

Depending on the type of link-up between the steelworks and rolling mill, distinction is made between cold charging, hot charging and direct rolling of the billets in the rolling mill (Fig. 5).

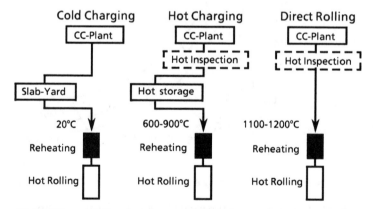

Fig. 5 Material flow between continuous caster and rolling mill.

In the case of cold charging, the moving stock is inductively heated up from ambient temperature within a minimum of time. In the case of hot charging and direct rolling, the casting heat is utilised for the rolling process. In the event of hot charging, the billets can be kept at a temperature between 600 °C and 900 °C (1112 °F and 1652 °F) in a holding facility to preserve the casting heat. To optimize steel quality the billets are brought up to the required rolling temperature within a minimum of time by inductive heating. Dwell times at only 600 °C - 900 °C (1112 °F and 1652 °F), in conjunction with rapid additional heating by induction, reduce surface decarburisation and scale formation. Coarse grain formation and intercrystalline corrosion are obviated.

In the case of direct rolling, the casting heat is fully utilised. The continuous caster is directly linked up with the rolling mill. Any temperature losses are compensated by induction heating.

In practice a mixed operating mode will result (Fig. 6), i.e. cold charging as well as hot charging or direct rolling must be possible.

In the event of direct rolling, the billets go direct to the rolling mill and are inductively reheated en route. If the rolling mill cannot process all the arriving billets, the surplus goes to a holding facility and is processed as required. Billets not required for direct rolling, i.e. when the holding facility is full, are allowed to cool to ambient temperature in the billet store and must subsequently be reheated. This reheating can be done by induction. The production capacity of this heater depends on the portion of cold charge within the total production. The production capacity of the heater for this duty is usually considerably lower than the total production capacity. Preheated billets are taken over directly from the induction reheater or run first through the holding facility, depending on the production sequence.

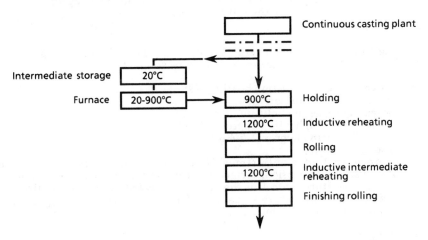

Fig. 6 Material flow with hot/cold charging.

Induction Heating in Rolling Mills

Application of induction heating is possible in the area in front of or within the rolling mill (Fig.7). The aim and objective is always to achieve optimum temperature control for the rolling process. If steelwork and rolling mill are not linked up, the requirement is that the billets must be heated from ambient temperature to the optimum rolling temperature for the material grade involved in front of the roughing train.

During the rolling process the requirement is to selectively compensate any temperature losses which may occur. These can consist of a general temperature drop over the full billet length or differing temperature losses at the head and tail ends of very long billets in the event of long handling distances. In this case, induction heating affords substantial advantages. Where steelwork and rolling mill are not directly linked up, the requirement that all billets have to be heated from ambient temperature results immediately.

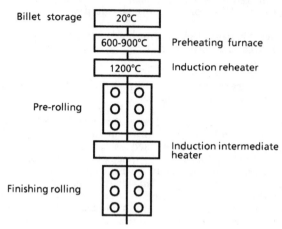

Fig. 7 Temperature control by induction heating
in the rolling process.

Depending on the structure of energy supply costs, the optimum concept providing for minimum heating costs and best technology must be adopted.

Advantages afforded by induction heating are:
- low surface decarburisation,
- low scaling losses, namely,
  with induction heating approx. 0.3-0.5 %,
  with fuel firing approx. 1.5-4.0 %,
- no adhering scale,
- no uncontrolled dwell times in the furnace, hence no overheating, intercrystalline corrosion, surface decarburisation and scale formation,
- exact temperature control, as the heat is generated where it is required,
- flexibility in temperature adjustment for the most varied steel grades- even for the production of small lot sizes.

Disadvantages:
- Power costs are considerably higher than the costs of gas, depending on the district.

Combination of Fuel-Fired Furnace + Induction Heating System

If the billets are preheated to 800 - 900 °C (1472 - 1652 °F) in a fuel-fired furnace, they can remain there for an extended period without suffering any appreciable metallurgical damage. All steel grades can be heated up to this preselected temperature range. An induction heater is then used for heating to rolling temperature with the aforementioned advantages. Even for small lots, different rolling temperatures (depending on the steel grade) can be achieved within a minimum of time. (Fig.8 shows scale formation as a function of temperature).

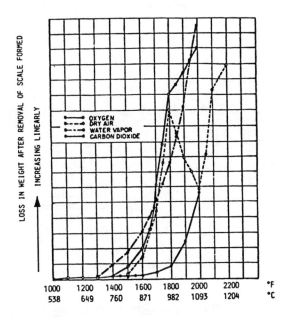

Fig. 8 Comparative effects of temperature and atmosphere
on the scaling of plain carbon steel, all exposed for a
constant time.

Induction Heating from Ambient to Rolling Temperature

In the case of smaller throughputs, up to approximately 25t/h, billets are heated completely in an induction heater. Heating is effected on line with the rolling mill (Fig.9). Such a heating plant can consist of the following groups:

- Depository magazine with billet separation
- Weighing facility for recording the individual billet weights. This information is automatically transmitted to the rolling mill control station
- Run-in roller table to convey the billets into the induction heater
- Collecting trough for returned heated billets
- Heating section with inductive type heat-up zone and resistance heated soaking zone

Sequence of operations:

Billets are individually conveyed from the magazine into a weighing facility. The weights etc. are transmitted to the rolling mill control. After weighing, the individual billets are automatically transferred to the run-in roller table of the heater line. The billets are now continuously brought up to rolling temperature as they move through an induction type heating section. The heater throughput corresponds to the mean throughput of the rolling mill. To ensure equal temperatures at the head and tail ends, the billets then enter a resistance heated soaking zone. From there they are fed into the rolling mill at the required entry speed (which may be considerably higher than the heater feed rate). Direct coupling of the heating section and the rolling mill results in least possible temperature losses at the billet surface.

In the event of any trouble in the rolling mill it is always possible to convey a heated billet in reverse direction out of the heating line and deposit it in a separate trough for cooling.

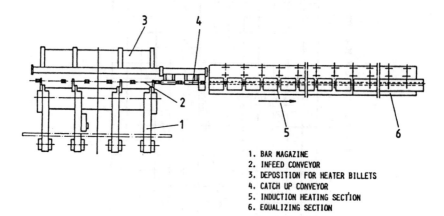

1. BAR MAGAZINE
2. INFEED CONVEYOR
3. DEPOSITION FOR HEATER BILLETS
4. CATCH UP CONVEYOR
5. INDUCTION HEATING SECTION
6. EQUALIZING SECTION

Fig.9 Induction heater with bar magazine.

Induction Heating to Enhance Temperature Control During the Rolling Process

Another application of inductive heating is selective compensation of thermal losses resulting from radiation during transport (Fig.10). The problem in the following example is that, due to excessive thermal losses at the surface, the elongation of core and surface differ so much that surface cracks result.

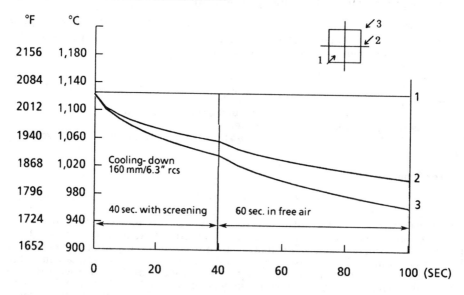

Fig.10   Cooling curve of a billet 160mm/6.3in rcs.

In the case under consideration the cross-section to be heated is 160mm (6.3in) rcs. Due to cooling in the section from the furnace up to the entry into the heater, the surface is up to 100°C (180°F) colder than the core. The core temperature remains constant.

By selective reheating (Fig.11) the surface temperature can be brought up to the core temperature or even to higher level. Energy consumption is approx. 11-15 kWh/t.

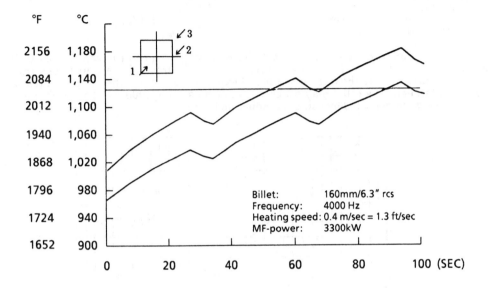

Fig.11   Temperature profile of indirect reheating.

Intermediate Heating Behind a Reversing Stand

Fig. 12 is an example of an intermediate heating system located behind a reversing stand.

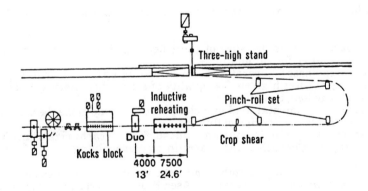

Fig. 12   Induction reheating system between roughing mill and continuous mill.

Billets up to 85m (279ft) long run through a closed channel from the reversing stand to the continuous train. In this instance the billet cross-section is flat-oval 61×35.5mm (2.4×1.4in).

Technical data are:

MF rating	2300kW
Frequency	3000Hz
Maximum heating temperature	1200°C (2190°F)
Temperature differences	30°C (54°F)

The maximum conveying speed from the reversing stand to the continuous train is maximum 4m/s (13ft/s). The minimum entry speed in front of the continuous train is 0.5m/s (1.6ft/s). Due to the differing times of exposure to free air, the temperatures of the head and tail ends of a billet may differ by as much as 300°C (900°F). This difference can be selectively compensated by using an induction heater (Fig. 13)

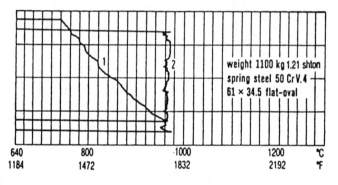

Fig. 13   Temperature vs. time characteristic of a billet 85m (279 ft) long, made of 50CrV4.

Summary

Induction heating systems of Asea Brown Boveri are characterized by the following advantages:

- Short heating sections (because of short heating times) allow a compact design which is well suited for automation
- Immediate readiness for operation because heating-up times are not required
- Little scaling ensures high surface quality of the product, narrow tolerances in the dimensions, and long tool lives
- Minimized environmental load and clean working conditions, because of low heat load and complete elimination of flue gas
- Reduced coarse grain formation and surface decarburization, facilitating subsequent processes such as thermal refining or machining

The examples described in the rolling mill show numerous applications of induction billet heating. Other fields of application for Asea Brown Boveri heating systems are:

- Induction heating in thin slab production
- Induction heating in strip rolling mills
- Induction edge heating for thin slabs and strips
- Induction heating for strip coating

Literature

1. Fasholz Jörg, Induktive Erwärmung (Inductive heating), Physikalische Grundlagen und technische Anwendungen, Energie-Verlag GmbH, 6900 Heidelberg.

2. Grulke Norbert, Schmiedeblock-Erwärmungsanlagen in Kompaktausführung (Forging billet heaters of compact design), Brown-Boveri-Nachrichten, Jahrgang '58, Heft 1/1976, Seite 3 bis 11.

3. Annen Walter, Die induktive Nachwärmung von Knüppeln in Feineisen und Drahtstraßen (Induction post-heating of billets in light section iron and wire rod mills), BBC-Sonderdruck CH-IW 12 02 50 D.

4. Dötsch Erwing und Jürgen Heinz, BBC Dortmund, Induktive Zwischenerwärmung im Walzwerk (Intermidiate induction heating in the rolling mill), Vortrag VDEH-Walzwerksausschuß am 6.2.86 in Hagen.

DIRECT ROLLING AND INDUCTION FOR TEMPERATURE RESTORATION

Gerald J. Jackson

Ajax Magnethermic Corporation
Warren, Ohio

Control of steel quality in the casting practice and coordination of the melt shop and rolling mill are the two main prerequisites to allow operation of a plant as a direct rolling system. The benefits of direct rolling are energy savings, labor savings, capital equipment savings, and yield increases which all result in higher profits.

As casting technology progresses, direct rolling will likely see widespread use through the steel industry. Today in the United States, there are three installations operating an induction reheat line direct rolling off a caster. These are located at Quanex, Mac Steel Division in Jackson, Michigan; Nucor Steel Division in Darlington, South Carolina; and Nucor Steel in Norfolk, Nebraska.

The advantage that induction heating offers a direct rolling operation can only be fully understood after a brief discussion of the thermal problems encountered when a caster is coupled to a rolling mill. To increase overall yields when rolling bars, bar lengths off the caster are typically in the 30 to 40 foot range. This results in a significant temperature difference between the lead and tail ends of the bar as it leaves the caster. See Figs. 1 and 2.

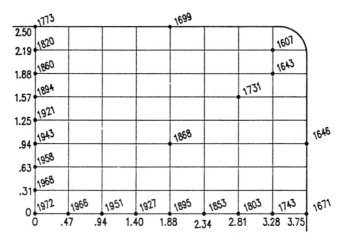

Fig. 1. Avg. Temp. $\sim$ 1807°F

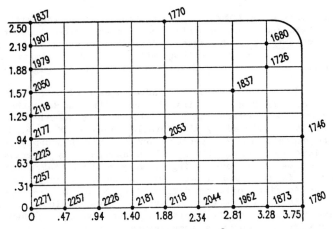

Fig. 2. Avg. Temp. ~ 1956°F

Also in a typical heat of 50 tons, the time to cast a ladle of steel would be 45 minutes to an hour which results in a reduction in the mean temperature of the steel as it leaves the caster as the heat is cast. Therefore, the heating system must satisfy the following requirements:

1. Must have the ability to taper heat the bars head to tail to compensate for the difference in time out of the mold.

2. Must have the ability to compensate for reduction in mean temperature from bar to bar during the time required to cast a heat.

3. Heating must be rapid to allow processing one bar at a time as it enters the rolling.

4. Space limitations usually dictate the heating time to be short, usually less than 25 feet.

The induction heating system installed at Nucor Steel's Darlington Plant is one example of a direct rolling system utilizing induction heating. Figure 3 is a schematic drawing of this system. This installation utilizes 6000 KW of power and four heating coils in a space of 12 feet prior to stand number one. The maximum instantaneous heating rate is 140 tons per hour (36.5 FPM on a 5" by 7½" bar) with a 250°F average temperature rise.

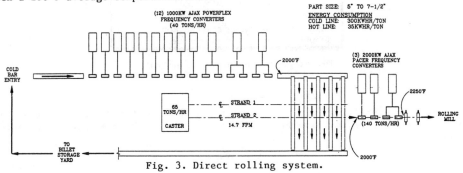

Fig. 3. Direct rolling system.

The energy consumption is 40 KW hours per ton based on the maximum temperature rise of 250°F. Bars enter the induction heater either from the caster or a reheat from cold. A key design consideration to optimizing a system like this is to have the caster, the heater, and the rolling mill in close proximity. This minimizes the cooling of the steel from the caster and makes for the most efficient system.

Induction heating bars provides two main advantages over other methods of heating that apply to direct rolling. The first is the ability to heat the steel directly to a depth below the surface. This is referred to as the depth of current penetration and is calculated by the formula shown in Fig. 4. For example, at 1000 hertz, the depth of current penetration is 5/8". This depth is the volume of the bar in which the current flows and as a result is directly heated. This characteristic allows induction heating to very rapidly heat a bar from a caster where the surface is the coldest area of the bar. Secondly, because the power delivered to the heating coils is developed by a static frequency changer, it responds very rapidly to the control system's instructions to raise or lower power. Allowing the power to be rapidly changed gives the induction heater the ability to take into account a temperature taper in the bar to be heated. Although it is possible to reheat bars from a caster in a gas furnace, many bars must be in process to allow sufficient soak time to produce a uniformly heated bar.

Depth of Current Penetration

Approximately 87% of heat generated in load is within the depth of current penetration. The depth of penetration is defined by the following formula:

$$d = 3160 \sqrt{\frac{\rho}{\mu f}}$$

d = depth ρ = resistivity of load

f = frequency Hz μ = permeability (1 for non-magnetic steel)

Heating Steel to 2250° F

60 hertz, d =	2.6"
180 hertz, d =	1.5"
300 hertz, d =	1.22"
1,000 hertz, d =	.64"
3,000 hertz, d =	.37"
10,000 hertz, d =	.20"

For good efficiency, the cross section should be about 3 times the d value.

Fig. 4. Depth of current penetration.

The temperature profiles below are a quarter grid of a 5" x 7½" carbon steel bar. The temperature profiles have been calculated using a finite element calculations technique.

A. Fig. 1:   Lead end of a 30 foot bar at the entry to the induction heater.
              Average temperature 1807°F.

B. Fig. 2:   Tail end of the same 30 foot bar at the entry to the induction
              heater.
              Average temperature 1956°F.

The higher average temperature at the tail end of the bar is due to the continuous
casting process and the greater time in air the lead end has to cool.

A. Fig. 5:   The lead end of the bar as it exits the induction heater.
              Average temperature 2214°F.

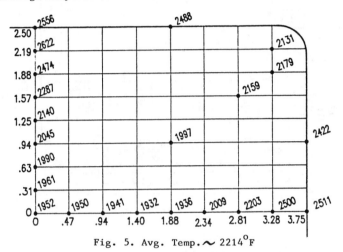

Fig. 5. Avg. Temp. ∼ 2214°F

B. Fig. 6:   The tail end of the bar as it exits the induction heater.
              Average temperature 2200°F.

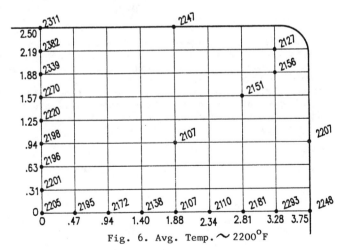

Fig. 6. Avg. Temp. ∼ 2200°F

The power developed in the bar is reduced automatically by the induction heater
control system (Fig. 7) to allow the heated bar delivered to the rolling mill to
have a uniform average temperature. The uniform average temperature is necessary
to prevent gauge problems on the mill as the bar is rolled.

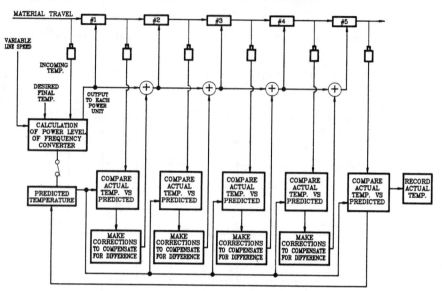

Fig. 7. Full auto control mode.

Because bars are heated individually, each bar receives the heat necessary to
achieve the desired result. When compared to many bars in a gas reheating furnace,
long heating times are required to achieve the same result. The one bar in process
allows the operator to quickly adjust the bar length at the caster shear to
optimize yield at the mill exit. Because of the short heating time with induction
(seconds vs. hours), there is almost no additional scale loss when compared with
the typical 2% scale loss of a gas furnace.

As a result of the complex temperature problems encountered when casting steel,
most companies today allow the steel to cool to ambient before reheating to roll.
For an economical comparison of the induction direct rolling operation vs. a
conventional gas furnace reheating, see Fig. 8. This comparison is based on 80% of
the steel cast being direct rolled with 20% reheated from cold with a second
induction heater. The gas furnace in this comparison is heating all bars from
ambient. The result is that when considering energy and scale loss, the induction
direct rolling system is approximately half the cost.

Heat Content at 2000°F = 631300 BTU's/Ton (184.8 KWHrs/Ton)
Heat Content at 2250°F = 703640 BTU's/Ton (206 KWHrs/Ton)
Natural Gas Rate: $4.00 Per MCF
Electric Rate: $0.045 Per KWHr
Annual Production: 230,000 Tons Per Year

          Assume 80% of 230,000 Tons per year can be direct
rolled off the caster.

| | TRADITIONAL GAS FURNACE (70-2250°F) | DIRECT ROLLING | | |
		INDUCTION HEATER (2000-2250°F)	INDUCTION HEATER (70-2000°F)	TOTAL
Tons/Yr.	230,000	230,000	46,000	230,000
Efficiency	50%	55%	65%	--
BTU'S/Ton	1,407,280	131,720	971,230	--
Energy Cost/Ton	$5.63	$1.74	$12.79	--
Scale Loss	2%	0.1%	0.25%	--
Scale Loss (Tons/Yr.)	4000	230	115	345
Scale Loss at $250/Ton	$1,000,000	$ 57,500	$ 28,750	$ 86,250
Energy Cost/Yr.	$1,294,900	$ 400,200	$ 588,710	$ 988,910
Total Cost/Yr.	$2,294,900	$ 457,700	$ 617,460	$1,075,160
Average Cost/Ton	$9.98	$1.99	$13.42	$4.67
Savings/Yr.	--	--	--	--

Fig. 8. Direct rolling preliminary cost study.

Labor savings and capital equipment costs are also areas where the induction heating system offers advantages. No dedicated operator is usually needed because the automatic control system operates the heater. Also because of the small space requirements and inline operation, handling equipment is eliminated.

Optimization of Enthalpy for Continuously Cast Strands in Hot Charging Applications

B. Lally[1], H. Henein[1] and L. Biegler[2]
[1]Department of Metallurgical Engineering and Materials Science
[2]Department of Chemical Engineering
Carnegie Mellon University
Pittsburgh, Pennsylvania 15213

Hot charging of continuously cast strands to hot rolling operations requires accurate control of strand enthalpy and quality. In this work mathematical optimization techniques and process models are used to determine process operating parameters that maximize strand enthalpy subject to constraints on strand quality that ensure a defect free cast strand. The models are based on a heat transfer analysis. Process parameters considered include the casting rate and the water spray settings for secondary cooling zones. Optimization problems representing billets and slabs are solved. In slabs, maximum enthalpy is found to occur at less than maximum casting rates.

Keywords

hot charging, continuous casting, billet, slab, optimization, modelling

Introduction

Continuous casting accounts for more than 50% of steel manufactured[1] and yields an 80% savings in energy over competing ingot processes.[2] With this impact on the steel market, it is especially worthwhile to develop optimization strategies for continuous casting processes. Additional energy savings can be realized through the use of direct charging strategies, where the cast strand is sent directly from the caster to the rolling mill while still hot, significantly reducing the cost of reheating the strand before rolling. In order to successfully practice hot charging a number of coordinated strategies must be implemented[3, 4, 5, 6] for steelmaking and caster operation, for maximization and control of cast strand temperature, for efficient and fast transport of hot strands to the rolling mill and for optimized sequencing[7] and consolidation[8] of operations from steelmaking to rolling. In practice these strategies include precise temperature and chemistry control of molten steel, caster breakout detection, mold level control and heat conservation of the strand during casting and during transport to the rolling mill. Defect tracking is clearly a major issue to address in implementating a hot charging practice. The quality of the strand must be assessed quickly on line (i.e., while the strand is being cast). An alternate approach is a defect free practice. This involves operating at process variables that ensure the optimized utilization of equipment at minimum cost while producing a defect free product. Furthermore, a comprehensive sensor network is required to measure the current operating status and to detect deviations from normal practice.[9] This paper outlines a method for determining caster operating parameters that will yield the maximum amount of thermal energy that can be retained in the cast strand in preparation for direct charging. High quality cast product will be produced for a defect free practice.

The problem of maximizing the retained thermal energy subject to quality constraints comprises an optimization problem. The problem is nonlinear in nature, as the retained enthalpy and the quality constraint functions are nonlinearly related to the process control variables. A general method for determining optima in nonlinear optimization problems applied to continuous casting has been previously presented.[10, 11] The method is briefly summarized below.

Within the optimization framework, continuous casting is modelled in terms of heat transfer. The relationship between the objective and constraint functions and the process control variables is determined by solution of a nonlinear partial differential equation; hence, it is necessary to solve a nonlinear optimization problem. A numerical solution to the heat transfer equation is used as a model to predict the steady state temperature field in the cast strand that results from a given set of process variables.[11] This temperature field is then used to calculate the values of the objective and constraint functions required by the optimization procedure.

The optimization procedure is an iterative one, employing the Successive Quadratic Approximation (SQP) approach.[12, 13] The iterations are illustrated in Figure 1. Initial estimates of the process variables are chosen and the thermal profile in the strand (the process state) is calculated using the model. Objective and constraint functions are determined from the calculated state information, and returned to the optimizer. Information concerning the derivatives of the objective and constraint functions with respect to the process variables also need to be calculated. With this information the optimizer calculates a new trial point and the process continues until a set of optimality conditions is satisfied.

In this work, the optimization techniques are applied to the problem of determining the caster operating variables that result in maximizing the enthalpy in a cast strand (measured at the cutoff point). Strict process constraints are enforced to ensure casting of a quality product. The results are compared with previous results[11] that determine maximum casting rates and, for billet and slab casting, maximum enthalpy conditions are found to occur at casting speeds less than the maximum allowed casting speeds.

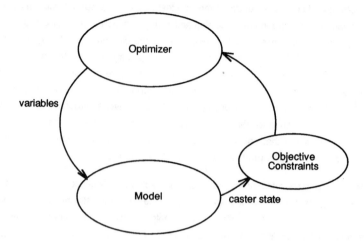

Figure 1: Pictorial representation of optimization iterations.

Optimization Problems

In this paper, the solutions to four continuous casting optimization problems are presented. The enthalpy in a transverse cross section (calculated as the section passes the cutoff station) has been maximized for the case of a billet caster operated at fixed and variable casting rates, and for a slab caster at variable casting rates. Some previously presented results concerning maximum permissible casting rates[10] are also included for comparison purposes. The optimization cases are summarized in Table 1.

Two models were used to represent continuous casters in this work. The billet caster was modelled using a mathematical model developed by the authors[11] based on alternating direction finite difference techniques (model 1). The slab caster problems were solved using this same model, and an additional model developed by the Allegheny Ludlum Steel Corporation[14] (model 2), based on the strongly implicit method of Stone.[15] Model 1 is a very general model capable of representing heat transfer in continuously cast sections ranging in size from small billets to large slabs. Model 2 was developed specifically for a particular slab caster. The two models were used for the slab caster problems so that the results from the different types of models could be compared.

The geometry of the casters are summarized in Table 2. The simulated billet caster is based on an industrial casting machine in use at Inland Steel. It is used to cast 0.18 m x 0.18 m billets, using a 0.61 m mold. There are 4 spray cooling zones, with independently controlled water sprays in each zone. Unbending of the curved strand occurs 16.7 m from the meniscus. The optimization variables are limited to values of the four heat transfer coefficients that represent strand cooling in each spray zone (problem 2), as well as the casting rate (remainder of problems).

The slab caster is modelled after a caster in commercial use at Allegheny Ludlum Steel Corporation

Table 1: Summary of the characteristics of the optimization problems studied.

problem	cast section	objective	note
1	billet	maximum rate	
2	billet	maximum enthalpy	fixed rate
3	billet	maximum enthalpy	variable rate
4	slab	maximum rate	model 1
5	slab	maximum rate	model 2
6	slab	maximum enthalpy	model 1
7	slab	maximum enthalpy	model 2

and has six independently controlled cooling zones. This machine is used to cast 1.32 m x 0.20 m stainless slabs. The optimization variables available are the casting rate and six heat transfer coefficients. The values specified for the heat transfer coefficients represent the effect of the cooling water at the midface of the strand - because of the design of the spray system, the heat transfer coefficients decrease in a complicated manner across the wide face of the slab. The effect is represented as a heat transfer coefficient applied at the center of the wide face of the strand and a correction factor applied to the heat transfer coefficient for positions away from the centerline. The correction factors are shown in Figure 2.

Table 2: Geometry of the continuous casters.

	billet caster	slab caster
section size	0.18 m x 0.18 m	1.32 m x 0.20 m
mold length	0.61 m	0.76 m
spray zone lengths		
zone 1	0.09 m	0.25 m
zone 2	0.38 m	1.37 m
zone 3	1.83 m	2.89 m
zone 4	2.44 m	4.11 m
zone 5	-	1.57 m
zone 6	-	1.57 m
unbending point	16.70 m	17.12 m

The thermophysical properties of the steel used in the billet and slab problems were chosen to approximate 1010 carbon steel and 304 stainless grades respectively. These are summarized in Table 3. For simplicity, the properties are assumed to be constant functions of temperature outside of the 2 phase region, and the heat of fusion is released linearly between the liquidus and solidus.

The billet maximum enthalpy optimization problem is described by equation 1. This problem statement details the objective and constraint functions used. The constraints have been developed fully in a previous paper[11] and are summarized after the problem statement.

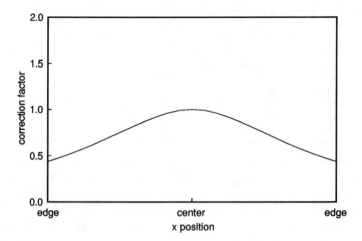

Figure 2: Values of the correction factor applied to the heat transfer coefficients for the wide face of the slab as a function of position.

Table 3: Thermal/physical properties of 1010 carbon steel used for the billet caster problems[16] and of 304 stainless steel used for the slab caster problems.[14]

	1010 carbon steel	304 stainless steel
solidus temperature	1477°C	1399°C
liquidus temperature	1522°C	1449°C
heat capacity		
solid	0.682 kJ/kg/°C	0.628 kJ/kg°C
liquid	0.682 kJ/kg/°C	0.712 kJ/kg°C
heat of fusion	272 kJ/kg	270 kJ/kg
thermal conductivity		
solid	0.0366 kW/m	0.0193 kW/m @ 650°C
		0.0343 kW/m @ 1399°C
liquid	0.0366 kW/m	0.100 kW/m
convection effect	7x	-
density		
solid	7400 kg/m^3	7500 kg/m^3
liquid	7700 kg/m^3	7500 kg/m^3
emissivity	0.8	0.8

$$\max \quad O(T(\mathbf{p}),\mathbf{p}) = \frac{\int_A T(x,y,Z_{cutoff})\,dx\,dy}{\int_A dx\,dy} \tag{1}$$

s.t.
$$0.01 \le p_1 \le 0.15 m/s$$
$$0.0 \le p_2 \le 2.0 kJ/m^2/s/^\circ C$$
$$0.0 \le p_3 \le 2.0 kJ/m^2/s/^\circ C$$
$$0.0 \le p_4 \le 2.0 kJ/m^2/s/^\circ C$$
$$0.0 \le p_5 \le 2.0 kJ/m^2/s/^\circ C$$
$$d_{shell} \ge 0.01 m$$
$$T(x_{center}, y_{center}, Z_{unbend}) \le 1477^\circ C$$
$$T_{i,j}^{reheat} \le 175^\circ C \qquad i=1,2,\ldots n_z+1 \qquad j=i,i+1,\ldots n_z+1$$
$$T_i^{max} \le 1200^\circ C \qquad i=1,2,\ldots n_z+1$$
$$T_{unbend} \ge 900^\circ C$$

The optimization variables p_i consist of the casting rate (p_1) and heat transfer coefficients $(p_2 - p_{n_z+1})$ that represent the water spray intensities in each of the n_z spray zones. In the billet case $n_z = 4$. The objective function $O(T(\mathbf{p}),\mathbf{p})$ calculates the average temperature of the strand in a transverse cross section. The integrals are evaluated over the cross sectional area of the strand at the cut off point. Since the heat capacity is fairly constant over the temperature range normally found in this cross section (this section is required to be completely solid by a constraint, hence no effect of the heat of fusion is found in the calculation), the average temperature of this section is considered a good approximation to the enthalpy of the strand.* $T(\mathbf{p})$ is the calculated temperature field. d_{shell} is the solid shell thickness at the mold exit and T represents temperatures at various points in the strand.

The objective function (the enthalpy at the cutoff point) is maximized subject to the following constraints:

- The casting rate is bounded, with a minimum of 0.01 m/s and a maximum of 0.15 m/s (equation 2). These bounds were chosen to limit the optimization procedure - they never become active.

$$p_i^L \le p_i \le p_i^U \qquad\qquad i = 1 \tag{2}$$

- The heat transfer coefficients are also bounded, with minima of 0.0 and maxima of 2.0 kJ/m^2/sec/°C (equation 3).[17]

---

*If this approximation is not considered sufficiently accurate, the actual enthalpy values could be substituted in equation 1.

$$p_i^L \leq p_i \leq p_i^U \qquad\qquad\qquad i = 2,3,\ldots n_{n_z+1} \qquad\qquad (3)$$

- The shell thickness at the mold exit must be at least 0.01 m (equation 4).[17]

$$\left(\left| X_{surface} - x \right| \text{ s.t. } T(x, Y_{center}, Z_{mold}) = T_S \right) \geq d_{shell}^{min} \qquad\qquad (4)$$

$$\left(\left| Y_{surface} - y \right| \text{ s.t. } T(X_{center}, y, Z_{mold}) = T_S \right) \geq d_{shell}^{min}$$

- Solidification must be completed at a distance not greater than 16.7 m from the meniscus. This constraint is enforced by requiring the centerline temperature at this point to be less than the solidus temperature, 1477°C (equation 5).

$$T(X_{center}, Y_{center}, Z_{unbend}) \leq T_s \qquad\qquad (5)$$

- The maximum surface reheating allowed (as defined by equation 6) must be less than 175°C. In this equation, T_j^{max} is the maximum surface temperature in zone j and T_i^{min} is the minimum surface temperature in zone i. This value was chosen since it is the amount of reheat that the model predicted under the nominal operating conditions for the caster. It is believed to be representative of the billet caster.[17]

$$T_{reheat}^{max} \geq T_j^{max} - T_i^{min} \qquad\qquad i = 1,2,\ldots n_z+1, \ \ j = i,i+1,\ldots n_z+1 \qquad (6)$$

- The midface surface temperature must always be less than 1200°C outside of the mold (equation 7).

$$\max T(x, Y_{surface}, z) \leq T_{surf}^{max} \qquad z_i^{start} \leq z \leq z_i^{end} \qquad i = 1,2,\ldots n_z \qquad (7)$$

$$\max T(X_{surface}, y, z) \leq T_{surf}^{max} \qquad z_i^{start} \leq z \leq z_i^{end} \qquad i = 1,2,\ldots n_z$$

- The midface surface temperature 16.7 m from the meniscus must be not less than 900°C (equation 8).

$$T(X_{surface}, y, Z_{unbend}) \geq T_{unbend}^{min} \qquad\qquad (8)$$

$$T(x, Y_{surface}, Z_{unbend}) \geq T_{unbend}^{min}$$

The formulation of the objective for the maximum rate problem is simpler, and is given by

$$\max \ p_1 \qquad\qquad (9)$$

The optimization problem for slabs is slightly different, as the slab caster has 6 cooling zones instead of 4, the casting rate is allowed to vary and the constraints have slightly different values. This problem is represented by equation 10.

$$\max \quad \frac{\int_A T(x,y,Z_{cutoff})\,dx\,dy}{\int_A dx\,dy} \tag{10}$$

s.t. $0.01 \leq p_1 \leq 0.15 m/s$

$0.0 \leq p_2 \leq 5.4 kJ/m^2/s/^{\circ}C$

$0.0 \leq p_3 \leq 5.4 kJ/m^2/s/^{\circ}C$

$0.0 \leq p_4 \leq 5.4 kJ/m^2/s/^{\circ}C$

$0.0 \leq p_5 \leq 5.4 kJ/m^2/s/^{\circ}C$

$0.0 \leq p_6 \leq 5.4 kJ/m^2/s/^{\circ}C$

$0.0 \leq p_7 \leq 5.4 kJ/m^2/s/^{\circ}C$

$d_{shell} \geq 0.01 m$

$T(x_{center}, y_{center}, z_{unbend}) \leq 1399^{\circ}C$

$T_{i,j}^{reheat} \leq 175^{\circ}C \qquad\qquad i=1,2, \dots n_z+1 \qquad j=i, i+1, \dots n_z+1$

$T_i^{max} \leq 1200^{\circ}C \qquad\qquad i=1,2, \dots n_z+1$

$T_{unbend} \geq 900^{\circ}C$

Discussion

The first solution presented in this paper concerns maximizing the enthalpy of a billet at a constant casting rate. Changes in casting rate have a direct influence on the average temperature at any fixed point along the strand, with increasing casting rate resulting in increasing enthalpy, hence the casting rate was held fixed at the normal operating practice of 0.03 m/s to isolate the effect that the secondary cooling system has on the average temperature of a transverse cross section. This casting rate value was found in previous work to be within the permissible upper and lower limits for this caster.[10] The initial and final states of the optimization trial are summarized in Table 4 (columns 1 and 2 respectively). During the solution of this problem, the average temperature in a transverse section of the strand at the cutoff point increased from 1147°C to 1196°C, an increase of 4.3%. The increase in objective function represents energy that does not need to be added to the cast piece in a hot charging situation. It is stressed that only the secondary cooling parameters were altered, and the constraints that represent quality were all satisfied.

The progress of the optimization process is shown in Figure 3, which iş a plot of the objective as a function of iterations. This optimization required 10 iterations, but the gains were essentially complete after 4. The greatest value of the objective is found after the second iteration; the operating variables used for this trial point result in violated constraints. The optimizer must reduce the objective function slightly in order to satisfy the constraints.

Table 4: Initial and final states for the billet maximum rate and enthalpy problems.

	initial state	max enthalpy (fixed rate)	max rate	max enthalpy (variable rate)
objective	1147	1196	1173	1200
casting rate	0.030 m/s	0.030 m/s	0.0326 m/s	0.031 m/s
heat transfer coefficients $(kJ/m^2/s/°C)$				
zone 1	0.900	0.838	1.211	0.695
zone 2	0.600	0.414	0.700	0.631
zone 3	0.400	0.295	0.417	0.355
zone 4	0.350	0.189	0.388	0.211

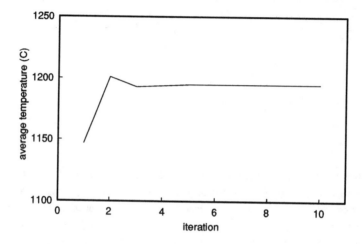

Figure 3: Objective function (average strand temperature of transverse cross section at cutoff point) as a function of SQP iterations for the billet maximum enthalpy problem at a fixed casting rate.

The increase in thermal energy provided by the optimization process is shown in Figures 4a and 4b. The figures depict the temperature field in the transverse cross section at the cutoff point. This comparison clearly shows the increase in thermal energy in the final state (note the increased diameters of the corresponding isotherms). It is quite evident that the thermal gradients in the strand cross section have also been reduced. Furthermore, these temperature values are clearly higher than those required for direct rolling and hot charging applications.[4, 6, 18]

In order to conserve as much enthalpy as possible for direct charging applications, it has been suggested that the caster should be operated at the maximum casting rate and the secondary cooling system adjusted accordingly.[3, 5, 6] The billet maximum enthalpy problem has been solved with a variable

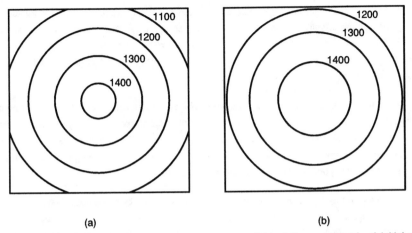

(a)   (b)

Figure 4: Temperature isotherms in the transverse cross sections of the cast strand. a) initial point of the optimization trial and b) final point. Temperatures are in degrees C.

casting rate to determine if casting at the maximum casting rate yields maximum strand enthalpy. The variables resulting in the maximum casting rate for the billet problem (summarized in column 3 of Table 4) have previously been determined[10] and are used to make the comparison.

The variables resulting from the solution of the maximum enthalpy problem with variable casting rate are tabulated in column 4 of Table 4. It is seen that the maximum enthalpy occurs at a casting rate that is less than the maximum casting rate and yields an increase in strand enthalpy of about 2%. This surprising effect is caused by the complex interactions that the control variables have on the objective and constraint functions. The heat transfer coefficients required to push the casting rate to a maximum (while satisfying the constraints) are significantly different from those used to maximize the enthalpy. Even with the greater enthalpy that would result from the increased casting rate, the heat transfer coefficients in the maximum rate case cause enough additional heat to be removed from the strand to reduce the overall heat content of the strand at the cutoff point. Note the larger heat transfer coefficients used to satisfy the constraints in the maximum rate case. This is a result that is not intuitively obvious, and would be difficult to discover by examination or trial and error.

The result that the casting rate in the maximum enthalpy solution is less than the maximum casting rate can be explained in another way. At the solution point to the maximum rate problem, the average temperature function is more sensitive to changes in the heat transfer coefficients than to changes in the casting rate. The water flow rates (represented by the heat transfer coefficients) at this point are large, in order to cause enough cooling to satisfy the metallurgical length constraint. Reductions in casting rate reduce the metallurgical length, and can be accompanied by reductions in cooling water, which increases the metallurgical length. In this case, the reductions in cooling water increase the strand average temperature more than the reduction in rate decreases it. The reheat and surface temperature constraints are involved in similar, albeit more complicated, trade offs. At first glance, it may appear that operating at maximum enthalpy will result in loss of productivity. However, considering increased energy savings and increased reheat furnace capacity,[6, 19] the net result could readily be reduced costs and greater rolling productivity. The break-even point between caster productivity and reduced energy costs will depend heavily on local mill conditions.

The maximum enthalpy problem with variable casting rate has also been solved for a slab caster. The problem is characterized by equation 10. This optimization problem is slightly larger than the billet problem, as the number of cooling zones has been increased to 6. This problem has been solved using the two different mathematical models. The initial state and the states resulting in maximum rate and maximum enthalpy using both models are summarized in Table 5.

The starting point for the optimization trials was selected as being representative of the normal operating conditions for the slab caster. According to both models these conditions resulted in reheat values approaching 300°C. This makes the starting point an infeasible point in the slab casting optimization problems. This large reheat value should not be interpreted as meaning that the normal shop practice produces steel of poor quality, but rather that the actual values for constraints applied to various machines and steel grades are not accurately known. 300°C reheat or more may be allowable in the casting of stainless grades.

The solution to this optimization problem, using model 1, required 37 SQP iterations, although the process was essentially finished after 15. The objective function and optimization parameter values at the initial point, at iteration 15 and at the final point are shown in Table 5. A comparison of the changes in these values from the initial point to iteration 15 and the changes from iteration 15 to the final point, along with the progress of the objective function shown in Figure 5 shows that only minor refinement occurs after iteration 15. Also, at iteration 15, all of the constraints are satisfied. The optimization process continues for 22 iterations beyond this point in an attempt to satisfy a termination tolerance that was purposely set to a very small value. The small value was chosen to ensure that the optimization process stopped at a true local extremum, and not at a nearly level surface or irregularity in the Lagrangian surface. Significant gains in heat content were made during this optimization trial. The objective function increased 16.3%, and the solution point was feasible, while the initial point was not. The surface reheating along the strand has been decreased nearly 125°C. The capability to improve the objective function and equilibrate the constraints simultaneously is an important feature of the method.

The maximum enthalpy problem was also solved using model 2. This model was substituted into the optimization framework to demonstrate the modularity of the model/optimization system and to compare optimization results determined from the different models. This modularity is an important factor in the application of this work to industrial applications, where accurate process models for specific processes or installations may already exist.

As in the previous trial, the problem was started from an infeasible initial state. The final state for

Table 5: Initial and final states for the slab maximum rate and enthalpy problems with model 1 and model 2. Also shown are the conditions at iteration 15 for model 1, where the optimization is essentially complete.

model 1

	initial	maximum rate	maximum enthalpy	
			iteration 15	final
objective	1018	-	1184	1184
casting rate	0.0144 m/s	0.0231 m/s	0.0220 m/s	0.0220 m/s
heat transfer coefficients				
$(kJ/m^2/s/°C)$				
zone 1	0.431	0.497	0.761	0.800
zone 2	0.642	0.758	0.213	0.208
zone 3	0.142	0.365	0.416	0.427
zone 4	0.049	0.198	0.166	0.172
zone 5	0.047	0.133	0.150	0.140
zone 6	0.052	0.105	0.077	0.068

model 2

	initial	maximum rate	maximum enthalpy
objective	1018	-	1184
casting rate	0.0144 m/s	0.0317 m/s	0.0245 m/s
heat transfer coefficients			
$(kJ/m^2/s/°C)$			
zone 1	0.431	0.370	1.066
zone 2	0.642	1.188	0.303
zone 3	0.142	0.719	0.579
zone 4	0.049	0.454	0.162
zone 5	0.047	0.340	0.001
zone 6	0.052	0.285	0.000

the maximum enthalpy problem agrees very closely with the final state predicted by the optimization problem using the model developed in this study. The parameter and objective function values are shown in Table 5. The objective function values agree almost exactly, while the operating parameters are quantitatively very close, and very similar qualitatively. The solution required 43 iterations to satisfy the very tight termination tolerance required, but as can be seen in Figure 6, the process was practically finished in 9 iterations.

In the slab maximum enthalpy problems, as in the billet enthalpy problems, the maximum enthalpy was found to occur at a casting rate that is less than the maximum allowed casting rate. Again, this is explained by the additional cooling water required to meet the constraints at the maximum casting rate. By slightly reducing the casting rate, the cooling water requirements can be reduced by an amount great enough to increase the enthalpy in the strand.

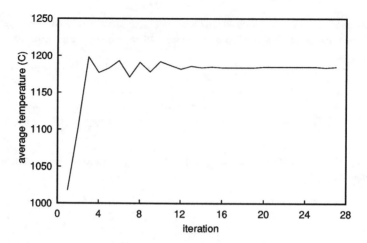

Figure 5: Objective function (average temperature in transverse cross section) as a function of SQP iterations for the slab maximum enthalpy problem with model 1.

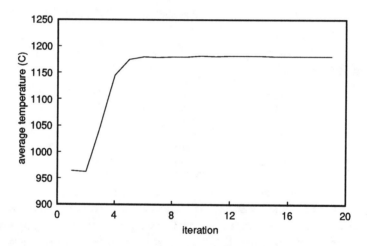

Figure 6: Objective function (average temperature) as a function of SQP iterations for the slab maximum enthalpy problem with model 2.

The differences in the solutions found using the two models is a result of differences in the mold boundary conditions. Model 2 predicts a wider range of allowable casting rates (greater maximum and lesser minimum) than model 1. This is caused by the mold boundary conditions. The model 2 mold boundary conditions have been constructed so as to increase the heat transfer coefficient (hence the heat transfer) in the mold with increasing casting rate. Model 1 uses a fixed heat transfer coefficient that is independent of casting rate. The variable heat transfer coefficient results in the mold being more effective at the maximum casting rate and allows the optimizer to select a greater casting rate when using model 2 than with model 1. The minimum casting rate is also less for model 2 than for model 1, as the mold becomes less effective at reduced casting rates. Because of this difference, the temperature profiles in the mold predicted by the two models agree very well at the initial casting rate (the rate at which the boundary condition for model 1 was determined), but show differences at other rates. The temperature profile at the mold exit can have a great effect on the reheat values found in the simulation, and this effect is made apparent in the different maximum casting rates determined. Heat transfer in the mold is found to have a large effect on the results. Accurate determination of mold boundary conditions is an extremely difficult experimental problem, and is currently impossible from a first principles viewpoint. This is an important area for further study in continuous caster modelling.

Summary and Conclusions

A mathematical model of heat transfer in continuous casting was constructed and used to calculate process operating states. The operating variables were manipulated using a sophisticated optimization algorithm to simultaneously determine a process state that satisfied constraints on the process and maximized a rating function that described internal heat content in the cast product. Operating conditions resulting in maximum strand enthalpy at the cut off point were found.

For both billets and slabs, the casting rate at which the strand enthalpy is maximized has been found to be less than the maximum casting rate. This is the case even though an increase in casting rate, holding all other casting parameters constant, leads to an increase in enthalpy. This seemingly contradictory situation is a result of the additional cooling water that must be used to satisfy the constraints as the casting rate is increased. This has been verified for both billet and slab casters, and for the case of slab casters, using two different mathematical models. This unexpected result is caused by the complex interactions of the casting variables and the objective and constraint functions. Heat transfer in the mold is found to have a large effect on the casting parameters chosen for defect free casting.

Acknowledgements

The authors wish to acknowledge Carnegie Mellon University, the Center for Iron and Steelmaking Research, its member companies and the National Science Foundation (grant 84-21112) for support of this research. We are also grateful for numerous discussions with Don McMahon and John Nauman of Allegheny Ludlum and Ken Blazek and Ismael Saucedo of Inland Steel.

References

1.   Continuously Cast Steel, 1977-1986, Iron and Steelmaker, Data originally from the International Iron and Steel Institute.

2.   S. Mizoguchi, T. Ohashi and T. Saeki, "Continuous Casting of Steel", *Annual Review of Material Science*, 1981, p. 151.

3.   K. Yoshida, T. Kimura, T. Watanabe and Y. Alcai, *Steelmaking Conference Proceedings*, ISS-AIME, 1987, p. 231-235.

4.   P. Schittly and G. Smarzynski, *Steelmaking Conference Proceedings*, ISS-AIME, 1987, p. 237-247.

5.   H. F. Schrewe, F. P. Pleschiutschnigg and D. Lohse, in *Continuous Casting, Volume 4: Design and Operation,* BookCrafters, Chelsea, Michigan, 1988, p. 213-226.

6.   H. Wiesinger, G. Holleis, K. Schwaha and F. Hirschmanner, in *Continuous Casting, Volume 4: Design and Operation,* BookCrafters, Chelsea, Michigan, 1988, p. 227-255.

7.   B. Lally, L. Biegler and H. Henein, "A Model for Sequencing a Continuous Casting Operation to Minimize Costs", *Transactions of the Iron and Steel Society*, Vol. 9, 1988, p. 123-130.

8.   L. R. Woodyatt, *Steelmaking Conference Proceedings*, ISS-AIME, 1987, p. 277-279.

9.   J. R. Cook, D. F. Ellerbrock, T. R. Dishum, H. N. G. Wadley and D. M. Boyd, *Steelmaking Conference Proceedings*, ISS-AIME, 1987, p. 285-293.

10.   B. Lally, H. Henein and L. Biegler, "Prediction of Optimal Operating Parameters for Continuous Casting of Billets", *Mathematical Modelling of Materials Processing Operations*, J. Szekely, L. B. Hales, H. Henein, N. Jarret, K. Rajamani and I. Samarasekera, ed., AIME-TMS, Warrendale, PA, November 1987, p. 1055-1069.

11.   B. Lally, *A Method for Optimizing Continuous Steel Casting Processes Using Mathematical Heat Flow Models,* PhD dissertation, Carnegie Mellon University, January 1988.

12.   S. P. Han, "A Globally Convergent Method for Nonlinear Programming", *Journal of Optimization Theory and Applications*, July 1977, p. 297-309.

13.   M. J. D. Powell, "A Fast Algorithm for Nonlinearly Constrained Optimization Problems", *Dundee Conference on Numerical Analysis*, 1977, p. 144-157.

14.   J. Nauman, Allegheny Ludlum Corporation, personal communication, 1987.

15.   H. L. Stone, "Iterative Solution of Implicit Approximations of Multidimensional Partial Differential Equations", *SIAM Journal on Numerical Analysis*, Vol. 5, No. 3, September 1968, p. 530-558.

16.   Y. S. Touloukian, *Thermophysical Properties of High Temperature Solid Materials, Volume 3: Ferrous Alloys,* MacMillan Company, New York, 1967.

17.   I. Saucedo, Inland Steel Corporation, personal communication, 1986.

18.   N. Moritama, O. Tsubakihara, M. Okimori, E. Ikezaki and K. Isogami, *Steelmaking Conference Proceedings*, ISS-AIME, 1987, p. 249-255.

19.   A. Etienne, C. Van der Hove and J. P. Baumal, *Steelmaking Conference Proceedings*, ISS-AIME, 1987, p. 295-302.

ELECTROMAGNETIC TECHNOLOGY FOR CONTINUOUS
CASTING IN THE STEEL INDUSTRY

M.B. Cenanovic, H.A. Maureira, M.K.C. Ng
Ontario Hydro Research Division
Toronto, Ontario, Canada
J. Mulcahy, L. Beitelman
J. Mulcahy Enterprises Inc.
Whitby, Ontario, Canada

ABSTRACT

Steelmaking in minimills primarily involves the melting of scrap ferrous material in electric arc furnaces, followed by continuous casting of the molten steel into blooms or billets.

A new method of continuously casting steel in near-net shape products is being developed at Lake Ontario Steel Company (LASCO) by J. Mulcahy Enterprises Inc. Ontario Hydro is providing support for this project, through its expertise in the field of electromagnetic applications. The device which has been developed is a closed-head, radially pulsating mold, which will produce a better quality surface and internal steel structure, to allow in-line hot rolling.

KEYWORDS

Continuous steel casting; direct rolling; novel electromagnetic pulsating mold.

INTRODUCTION

Ontario Hydro encourages the efficient use of electricity and promotes the development of innovative ways of using electricity which will result in increased productivity and competitiveness of Ontario Industry. Energy management programs are in place to achieve these objectives. It is in this context that Ontario Hydro's Research Division is prepared to provide expertise in the field of electromagnetic applications (Cenanovic, 1973, 1981, 1983).

Electromagnetic technologies have contributed greatly to the impressive progress achieved in the continuous casting of metals in the past decade. Electromagnetic stirring (Birat and Chone, 1983), molten metal flow control (Ayata and Fujimoto, 1988), levitation casting (Lowry and Frost, 1983), magnetic mold (Getselev and Martynov, 1973), shaping and guiding liquid metal in thin strip casting (U.K. Patent Application, 1980) are all examples of the application of electromagnetic techniques either to upgrade existing continuous casting methods or to provide a base for the development of new ones.

This paper reports a new method of continuous casting of steel based upon a high power pulsed electromagnetic field technology being developed jointly by J. Mulcahy Enterprises Inc. (JME), Ontario Hydro Research Division (OHR) and Lake Ontario Steel Company (LASCO). A high power pulsed electromagnetic field is employed to obtain radial pulsation of a closed-head casting mold. High frequency pulsations result in a drastic reduction of the friction force between the solid shell and mold. Due to friction reduction it becomes possible to cast a small cross section strand, e.g. bar, with a speed compatible to that of a rolling mill. Therefore, a technological base is originated for bar manufacturing via an integrated continuous casting-rolling system.

The following sections provide a general description of such an integrated system for production of special quality bars, as well as development of the central component of this system.

NEW PROCESS

At present, bar and rod stocks are cast into billet and bloom sections by means of conventional or horizontal continuous casting. Both casting technologies have inherent drawbacks associated with a stationary mold which prevents them from casting near-net shape products or being integrated with rolling mills. A low strand withdrawal speed and the need for billet surface conditioning prior to further processing are major problems.

As solidification proceeds in the mold, the shell tends to stick to the mold wall and as a result, friction occurs as long as the strand is being withdrawn from the mold. This friction produces stress in the thin shell and limits the withdrawal speed to a level which will not cause shell rupture. A novel method of reducing friction in the mold has been proposed (Mulcahy, 1985). Instead of oscillating the mold in an axial direction, as is employed in conventional and some horizontal casting techniques, a radial pulsation of the mold wall is introduced. The high frequency radial pulsation is achieved by means of a high power pulsed electromagnetic field. This subject will be discussed in more detail later in the paper.

A high frequency radial pulsation of the mold will result in a marked reduction of friction during strand withdrawal and therefore provide a high-speed, smooth withdrawal. Withdrawal speeds of 6 m/min and greater will be compatible to the speed of a bar rolling mill. The surface of the as-cast product will be virtually defect-free and therefore suitable for further continuous processing.

Another advantage of the radial pulsating movement over axial oscillation of the mold is that the pulsating mold can be readily mounted on the tundish (or any other molten metal dispenser) in either vertical or horizontal arrangements. The coupling between tundish and mold is sufficiently tight to protect the liquid metal from reoxidation as it enters the closed-head radially pulsating mold. Operation of an integrated continuous bar casting-rolling system is illustrated by the schematic layout shown in Fig. 1.

Molten metal is supplied from the ladle to the tundish and then, to the radially pulsating mold. The metal flow is controlled by the speed of strand withdrawal, although a slide gate provides conditions for the casting start and emergency strand shut-off. The strand, withdrawn at a speed of 6 m/min or greater, is suitable for direct rolling without having to reheat it to compensate for temperature loss. For example, 4 to 5 minutes after the beginning of solidification, the surface temperature of a 64 mm diameter (2-1/2") bar cooled mainly by radiation will be approximately 1150 deg. C. However, an induction heater is installed before the rolling mill to compensate for possible temperature fluctuations which may occur at the start of casting or be caused by other reasons in the course of casting.

Roughing stands and a series of finishing mill stands will provide the facilities needed for production of bar from maximum 76 mm (3") to 12.7 mm (0.5") diameter. The bar, depending on its cross section size, may then be coiled or cut to a required length and bundled for shipment.

This type of integrated operation provides a tight control of the initial material, finished product and the manufacturing process. Such control is an important prerequisite in the production of bars for highly demanding applications, i.e. automotive forgings, etc.

Integrated bar manufacturing technology will provide substantial benefits as compared to the existing billet casting route. A comparison between the operating costs of manufacturing 51 mm diameter bar from the stock of 64 mm diameter cast bar and from 133 mm sq. billet is given in Table 1. The data presented in the table is based on the operating results of a three-strand billet caster with an output of approximately 6,000 tons/week and a two-strand pulsating mold bar caster with a projected output of approximately 3,000 tons/week.

A comparison has been made in the three main areas of operation, i.e. energy consumption, yield loss and productivity expressed in manhours/ton. As seen, bar casting would result in substantial benefits in all operating areas. The total operating cost savings are estimated to be approximately $ 37.00/ton which represents 10% saving of costs. The economical benefits will be enhanced by accounting for lower capital costs for the bar casting installation. More compact and lighter casting equipment, elimination of a reheating furnace, billet conditioning equipment, facilities and space for billet storage, etc. will substantially reduce the capital cost component per ton of finished product.

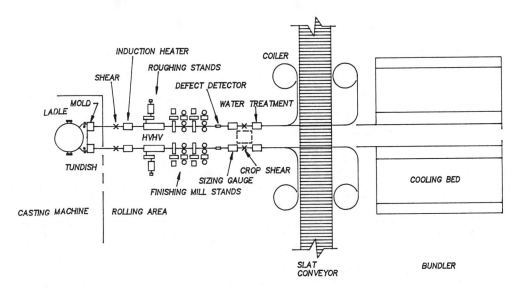

Fig. 1. Schematic Layout of Bar Casting/Rolling Mill

TABLE 1

Comparison of Operating Costs for
Manufacturing a 51 mm Bar from Billet Vs Cast Bar

I T E M S	BILLET 133 mm sq.	CAST BAR 64 mm diam.	RELATIVE COST SAVING (%)
1. ENERGY CONSUMPTION, KWh/T			
• REHEATING PRIOR TO ROLLING	454 (NATURAL GAS)	4.0 (ELECTRICITY)	95.5
• ENERGY RELATED TO ROLLING	13.2	2.9	78.0
TOTAL ENERGY	467.2	6.9	93.1
2. YIELD LOSSES, %			
• CASTER	3.7	< 1.0	73.0
• ROLLING MILL	4.2	< 1.0	76.2
TOTAL UNRECOVERABLE LOSS	7.9	< 2.0	74.7
• DOWNGRADED BILLETS/BARS	10	< 3.0	70.0
3. MANHOUR/TON	1.0	0.6	40.0

PRINCIPLE OF OPERATION

The radially pulsating mold concept is presently being evaluated mainly from the point of view of electromagnetic design and performance. Some work has also been done to establish mechanical design parameters. A schematic representation of the radially pulsating mold is shown in Fig. 2.

As seen, the arrangement includes two coaxial cylinders with a water gap between them and an electromagnetic coil. When the pulsed current passes through the coil wound over the outer cylinder or anvil, a pulsed magnetic field exerts radial pressure on the latter. This results in the radial elastic deflection of the anvil and thereby transmission of the pressure through the cooling water to the mold and ultimately to the steel cast. Under the applied pressure, elastic contraction of the mold's wall occurs.

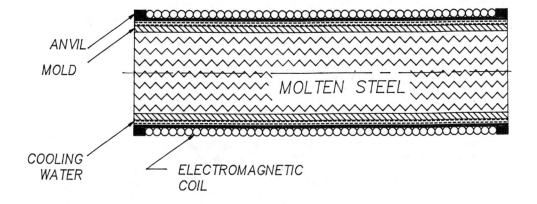

ANVIL

MOLD

COOLING
WATER

ELECTROMAGNETIC
COIL

MOLTEN STEEL

Fig. 2. Radially Pulsating Mold Concept

Once the pulse of magnetic pressure is nullified, the mold oscillates freely.
The magnetic pressure pulse is then re-applied after a long rest period and long
after any natural oscillation of the mold has died out. The concept is simple
and convenient, as it takes advantage of the presence of the cooling water and of
the outer cylinder (or anvil in this particular case) which is already part of
the assembly for the purpose of cooling the mold.

ELECTROMAGNETIC ANALYSIS

Pressure is applied to the anvil by means of electromagnetic coupling with a coil
which is wound around the anvil. The energy of a capacitor is discharged through
this coil such that a very high amplitude but short duration current pulse
results. An induced current is created in the anvil as a result of mutual
coupling between the coil and the anvil which is made of conductive material
(aluminum or copper). Electromagnetic force interaction causes the coil and the
anvil to repel each other. The coil is braced suitably to keep it from expanding
outwardly as this would result in a larger airgap between the coil and the anvil
which would in turn weaken the electromagnetic force interaction with the anvil.
The anvil is, on the other hand, free to respond and therefore its wall is driven
inwards. Energy efficiency considerations dictate that the above process must be
of very short duration. Furthermore a suitable switching scheme is used to allow
only half a cycle of natural oscillation of the electric circuit. In this way, a
significant amount of energy that remains in the capacitor is saved. A charging
DC source then restores the capacitor to its initial state of charge. The
magnetic pressure acting on the anvil equals the difference of magnetic energy
densities on both sides of the anvil's conductive wall as given by Eq. 1
(Cenanovic, 1983) in which

$$P = \frac{1}{2} \mu_o (H_o{}^2 - H_i{}^2) \qquad (1)$$

$\mu_o = 4\pi \times 10^{-7}$ is the permeability of free space in MKS units and H_o and H_i are the magnetic field intensities outside and inside the anvil. If the coil wound around the anvil is supplied with a sufficiently fast changing current, the magnetic field created by it will exist only in the region outside the anvil due to skin effect. Therefore, the anvil's wall thickness should be ideally not less than the skin penetration depth (which depends on the frequency of the time varying field and material properties) so that maximum magnetic pressure on the anvil from the outside will result.

SIMPLIFIED MECHANICAL ANALYSIS

For simplicity, the anvil is considered separately as a single component. Furthermore the assumption of an infinitely long, thin wall cylinder is made with uniform pressure applied in compression throughout its outer surface. Under these conditions, and also assuming no mechanical damping, the response of the anvil in breathing mode (radial pulsation) is described by a second order linear differential equation which is expressed in terms of b the cylinder's wall thickness, E the modulus of elasticity of the material, ρ the material density, R the undeformed radius, P the applied pressure and u the radial displacement (i.e. the change in radius $u = \Delta R$).

Pressure is applied to the anvil in pulses as shown by Fig. 3. The pulses are very narrow by comparison to their repetition period. Because of the long rest period between pulses it can be assumed that any pulsation of the anvil's wall dies out before the next pulse is applied. This is true in spite of the previous simplifying assumption of no mechanical damping. The response of the anvil to a single pressure pulse is therefore of interest. The response of the anvil was determined in terms of the displacement of the anvil's wall expressed as a function of the frequency of the applied pressure pulse. A normalized maximum displacement spectrum was then obtained (Fig. 4). The maximum displacement is normalized with respect to the static displacement (Eq. 2) and the frequency is expressed relative to the anvil's natural frequency (Eq. 3).

$$\left(\frac{\Delta R}{R} \right)_s = \frac{P_m}{E} \frac{R}{b} \tag{2}$$

$$\omega_n = \frac{1}{R} \sqrt{\frac{E}{\rho}} \tag{3}$$

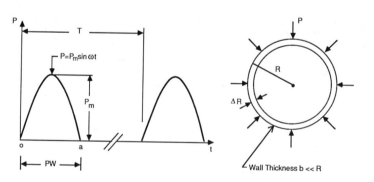

Fig. 3. Pressure Duty Cycle Applied to The Anvil

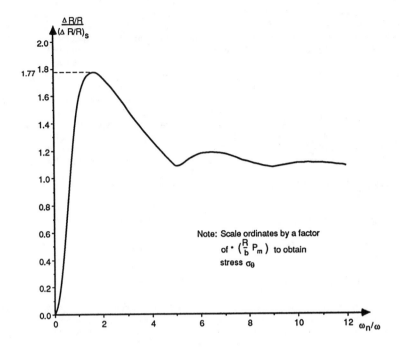

Fig. 4. Maximum Displacement Spectrum

The maximum hoop stress σ_θ is then obtained from the same normalized curve in Fig. 4 applying a scaling factor as indicated. This factor is derived from the relationship $\sigma_\theta = E\epsilon_\theta$ in which ϵ_θ is the circumferential relative displacement which in this case is equal to the radial relative displacement $\Delta R/R$.

CIRCUIT ANALYSIS

The magnetic coupling between the Anvil and the coil results in an equivalent inductance L seen from the terminals of the coil. L is given by Eq. 4 in terms of the coil's inductance L_{11}, Anvil's inductance L_{22} and mutual inductance L_{12}.

$$L = L_{11} - \frac{L_{12}^2}{L_{22}} \qquad (4)$$

The coil takes energy from a capacitor for half a cycle or alternatively a full cycle of natural oscillation of the electric circuit depending on the switching arrangement selected as part of the design of the power supply. To be specific and consistent with the previous mechanical analysis, a single pulse (i.e. half cycle) system is assumed.

The electrical parameters of interest are calculated using the usual equations of an L-C circuit. No damping may be assumed as the damping coefficient is expected to have a significantly low value (typically $\xi = 0.03$). Furthermore, no significant damping would occur because only the first half cycle of natural oscillation is let through. The effect of the circuit inductance or stray inductance due to supply cables, bus bar arrangement, etc. has to be considered however, as the load equivalent inductance L (from Eq. 9) is designed to have a low value. The low value of L is dictated by design trade-offs as explained below. The electrical parameters of interest can be readily calculated for selected values of capacitance and supply voltage (i.e. the initial DC voltage of the capacitor). The parameters of interest are the current pulse peak value and pulse width, and the peak magnetic energy stored in the anvil-coil arrangement. The peak magnetic pressure is then calculated as the peak magnetic energy density in the volume taken by the insulation between the work piece or anvil and the metal surface of the coil's copper conductor. In this work this is called the "air gap" by similarity with conventional magnetic equipment and devices. For simplicity, a uniform field distribution may be assumed in the air gap. The RMS value of the current by the coil in repetitive pulse operation is another parameter of interest and it is obtained from the peak value of the current pulse for a specified duty cycle. The current duty cycle is the same as the pressure duty cycle (Fig. 3) as defined by pulse width and repetition rate.

DESIGN CONSIDERATIONS

The main design trade offs are discussed below based on the results from the previous analyses.

1.   Anvil's Wall Thickness

For a thin cylinder the material frequency in breathing mode is independent of the wall thickness. On the other hand the static displacement increases in inverse proportion to the wall thickness. Therefore it is desirable to have a cylinder (anvil) as thin as possible provided that (a) the wall thickness is not less than the magnetic field penetration depth so the field H_i inside the anvil is null and the applied magnetic pressure is maximum (Eq. 1.) (b) the wall is still sufficiently thick to provide structural integrity for the expected stress levels σ_θ (from Fig. 4.).

2.   Copper Conductor Gauge

A thin conductor is more easily wound and takes less space. However, its thickness (assuming rectangular or square cross section is used) should also be ideally not less than the skin penetration depth so that most of the field is confined within the gap space between the bare part of the copper conductor and the anvil. This is to ensure a strong magnetic field H_o between the anvil and the coil hence the applied magnetic pressure. The gauge has to be also heavy enough for the expected RMS current and also heavy enough to yield a low resistance value (consequently a low damping coefficient).

3.   Pulse Width

The pressure pulse width is the same as the current pulse width which is one half the natural period of the L - C circuit. The pulse width is related to the frequency ω of sinusoidal variation of the pressure P which in turn determines the maximum (peak) displacement and stress according to the displacement (stress) spectrum. It is desirable to operate at a value of ω about 1.8 times less than the natural frequency ω_n (Eq. 3) so that the peak displacement is maximized. This is subject, however, to the limitation imposed by the power supply system in terms of stray inductance or circuit inductance L_s. L_s imposes a limit to the speed of the discharge of the L - C circuit hence pulse width (PW) and it also affects the net magnetic energy stored in the anvil's gap i.e. part of the magnetic energy is stored in the stray field and is therefore not useful to apply pressure on the anvil.   If the coil inductance L is increased significantly relative to L_s, the effect of L_s becomes less pronounced and eventually irrelevant, however, the pulse width PW would increase possibly too much and the anvil's peak displacement is minimized approaching the static displacement but more importantly the I^2R losses may become so high as to make the system unacceptably inefficient.

In light of the design trade offs outlined above the parameters of interest were calculated for a series of design options.   A particular design was chosen trying to keep the power supply requirements as low as possible for a desired level of performance.   A prototype consisting of an aluminum anvil with a coil of magnet wire wound around it was built and tested.   This prototype is essentially a thin wall cylinder approximately 25 cm long and 8 cm in diameter with expected wall deflection of about 125 micrometres.   A copper prototype anvil is also being built.   Copper has a higher modulus of elasticity than aluminum and should experience less displacement but has the advantage of better thermal conductivity.

TEST RESULTS

The prototype anvil was tested at Ontario Hydro's Electromagnetic Laboratory. The anvil's displacement was measured using a set of adjustable gap precision contacts that provide a close-open indication.   Good tests results have been obtained with the anvil by itself and a series of tests will be run with a complete mold assembly in the near future.   Metallurgical results will be known only after the complete mold assembly is tested and subject to trial runs in a steel casting plant.   The results obtained so far show a radial anvil's displacement of 50 to 200 micrometres at various points on the anvil's length with a sinusoidal current pulse of about 18 kA (peak) for a duration of approximately 100 microseconds.

The above test results closely confirm predicted performance of the anvil.   The maximum displacement should increase steadily between 25 and 150 microns when the load inductance increased between 1 and 6 μH.   A value of 2 μH close to the measured circuit (stray) inductance of 1.5 μH was used.   Higher inductance values were precluded by mechanical stress and thermal limits.   A ratio of anvil's natural frequency to frequency of the applied pressure pulse of 3.8 was used.   A lower value approaching 1.8 would be desirable according to Fig. 4.   This is not possible in practical terms because it would require a reduction of the capacitance and/or the inductance of the power supply.   Capacitances of 120 μF and 300 μF were used, the above test results correspond to 300 μF.   Lower capacitance values are restricted in order to keep the supply voltage down to a desired level of 2500 volts DC.   Lower inductance values are restricted in order to keep most of the energy supplied from the capacitor from being transferred to the stray magnetic field.

CONCLUSION

A joint effort between J. Mulcahy Enterprises Inc., Ontario Hydro and Lake Ontario Steel Company is underway to develop and employ a novel electrotechnique for continuous casting of steel.   It is anticipated that operating and techno-economical benefits associated with this technology will make it especially attractive for a low volume production of special quality bar.

REFERENCES

Ayata, K., and Fujimoto, T., Development of Electromagnetic Valve for Continuous Casting TUNDISH, Trans. ISIJ. Vol. 28, No. 1, p. 13 - 35. 1988.

Birat, J.P. and Chone J., Electromagnetic Stirring an Billet, Bloom and Slab Continuous Casters, State-of-the-Art in 1982, Ironmaking and Steelmaking. Vol. 10, No. 6, p. 269. 1983.

Cenanovic, M., Acceleration of the Flow in a Channel Furnace, Master of Applied Science Thesis, University of Toronto, 1973.

Cenanovic, M., Electromagnetic Forming of Zirconium and Other Strong Alloys, Proceedings of 20th Annual Conference of Metallurgists, Hamilton, Ontario, August 23-27, 1981.

Cenanovic, M., Magnetic Metal Forming by Reversed Electromagnetic Forces, Proceedings of 4th IEEE International Pulsed Power Conference, Albuquerque, New Mexico, Paper 6.4, June 1983.

Getselev, Z.N. and Martynov, G.I., Calculation of the Main Electromagnetic Parameters of an Apparatus for Ingot Shaping During Continuous Casting, Magnethydrodynamics, No 4, p 135-138, 1973.

Lowry, H.R. and Frost, R.T., Continuous Metal Casting Method, Apparatus and Product, U.S. Patent, No. 4, 414, 285, Nov. 8, 1983.

Mulcahy, J.A., Continuous Casting of Steel, US Patent No. 4,522,249, June 11, 1985.

British Steel Corporation, U.K. Patent Application No. 2, 048, 140. Dec. 10, 1980.

SESSION 3

SURFACE QUALITY AND SENSORS
FOR SURFACE QUALITY

Chairpersons: R. W. Pugh (Stelco)
G. Kamal (Ivaco)

COMPUTER AIDED QUALITY CONTROL FOR BLOOMS AND BILLETS

G. Holleis*, W. Dutzler*, L. Pochmarski**, H. Preissl*,
K. L. Schawha*
* Voest-Alpine Industrieanlagenbau, Linz, Austria
** Voest-Alpine Stahl, Donawitz, Austria

ABSTRACT

The principles of Computer Aided Quality Control (CAQC) and its application with
regards to continuous casting are described with particular emphasis on bloom
and billet casting. The concept of CAQC was developed by VOEST-ALPINE in
cooperation with BETHLEHEM STEEL CORP. for their slab casters at Sparrows Point
and Burns Harbor in 1983 and was put into operation in the first half of 1986.
Since this time, VOEST-ALPINE has implemented CAQC systems at POSCO's no. 3
slab caster in South Korea, ISCOR's slab caster in South Africa, and POSCO's
No. 1 slab caster and   at VOEST-ALPINE's slab caster No. 4 in Linz, Austria.

Depending on the purpose five CAQC variants of different complexity are available,
which will be of particular interest to bloom/billet casting. Whereas for POSCO's
bloom caster the most complex CAQC system has been realized, a system of somewhat
reduced complexity is presently under development for the bloom and billet casters
at VOEST-ALPINE's Donawitz works. The specific features of the two last
examples are discussed in the paper.

KEYWORDS

Computer aided quality control, Process Control, quality prediction, CC-production
disposition, strand tracking

INTRODUCTION

Quality control has become the key for successful and competitive steelmaking.
For this purpose statistical process control (SPC) has been widely introduced
and utilized to reduce deviations in operation and to provide a high degree
of consistency in product quality.

VOEST-ALPINE has developed a computer aided quality control system (CAQC)
which goes much beyond the point of quality control by SPC since CAQC calculates
the quality of each individual slab or bloom by complex correlation of many
process parameters rather than predicting quality levels from deviations.

The CAQC system has been successfully introduced on several slab casters i.e.
BETHLEHEM STEEL at Sparrows Point No. 1 and 2 and Burns Harbor No. 2,
POSCO No. 3/Korea, ISCOR/South Africa, POSCO No. 1/Korea and VOEST-ALPINE
No. 4. The system has also been adapted for bloom casting with different
levels of complexity depending on the final goals. For example at POSCO

bloom caster No. 1 the complex version from slab casting has been implemented. Presently a somewhat reduced CAQC package is realized for the bloom and billet caster at VOEST-ALPINE Donawitz works.

In the present paper the basic philosophy of CAQC is described and package solutions for different requirements are discussed with particular emphasis on bloom/billet casting.

Any CAQC system is based on the exhaustive metallurgical knowledge about the connex of cast and final product defects as well about of the dependence of defects on process parameters. Consequently, CAQC spans   the whole production route from steelmaking, ladle metallurgy, casting to rolling and has to be integrated into the whole production control system. Since details of the system have been published previously by Fastner et al (1984,1988) the various functions shall only briefly be discussed.

Specification of product quality

Figures 1 and 2 schematically show characteristic interior and surface defects of CC blooms, which constitute quality variables to be taken into account by CAQC. These defects are clearly defined and can be identified quantitatively in the as-cast product.

As an example the list of quality variables for slab casting is shown in fig. 3, which also gives the measuring methods and criteria for each quality variable. Consequently, product quality is defined by permissible limits for each quality variable , which is done via maximum allowable rating values.

Basic functions of CAQC (fig. 4)

Quality planning. From the demands on the final product regarding surface and internal conditions the required slab quality is planned automatically by using decision tables stored in the metallurgical data base.

Determination of casting practice and in-line correction (fig. 5). Prior to production the operating practice, suited to achieve the required slab quality, is determined for each production step. In the case of a deviation during production, the practice information is dynamically modified in order to provide compensation in a downstream production phase (in-line correction).

Process tracking (fig. 6). All important process data and disturbances relevant to quality are collected and related to the corresponding production sections. Segment lengths are typically between 500 and 1000 mm. A complete set of data is stored for each strand segment of a heat.

Quality Predictions (fig. 7). The quality of the individual slabs is calculated by means of metallurgical functions, which combine the tracked segment data into a quality result in the form of a quality code which contains rating values for every quality characteristic.

Disposition (fig. 8). The predicted and planned quality codes are compared. The correct disposition is derived from the deviations of the quality codes by using the action tables, so that further slab disposition e.g. hot charging, inspection, scarfing etc. is automatically determined.

Auxiliary functions of CAQC

Statistically evaluation and reports. The data collected from each strand segment contains extensive information about the produciton process. This is an invaluable source for statistical evaluations and reporting. Standard information and evaluations provide both a detailed insight into the various production phases and a means to increase metallurgical know-how and to optimize the data base of CAQC.

<u>Maintenance of metallurgical data base</u>. The metallurgical know-how needed for the CAQC functions is stored in the metallurgical data base in the form of rules and decision tables. The data base is structured to allow easy maintenance by means of a display and dialogue system. In this way, it supports the development and improvement of the system without programme changes.

Alternative solutions

To meet the different demands of quality control 5 different variants of CAQC are offered by VOEST-ALPINE. These variants of staggered complexity range from a simplified to a very sophisticated package as indicated in fig. 9 and discussed below:

Variant 1) Defect Tracking (fig. 10)
            - Process Tracking
            - Statistical evaluations and reports
Variant 2) Quality Tracking (fig. 10)
            - Process Tracking
            - Calculation of the deviation between the planned and actual
              caster practices
            - Statistical evaluations and reports
Variant 3) Standard Quality Prediction (fig. 11)
            - Process Tracking
            - Quality Prediction (standard)
              (2 states are distinguished 0 - 1 corresponding to
              go/no-go recommendation)
            - Statistical evaluations and reports
Variant 4) Advanced Quality Prediction (fig. 11)
            - Process Tracking
            - Quality Prediction (complex)
              (10 states are distinguished 0 - 9 ranging from small to high
              probability of defects)
            - Statistical evaluations and reports
Variant 5) Comprehensive Quality Control (fig. 12)
            - Determination of required product quality and operation practice
            - Process Tracking
            - Quality Prediction (complex) (10 states are distinguished,
              0 - 9 ranging from small to high probability of defects)
            - Slab Disposition
            - Statistical evaluations and reports

CAQC system at Posco bloom caster

As part of the revamping of the bloom caster at Pohang Works of Posco/South Korea, which included the change from a 3-strand to a 4-strand machine, the caster has been fully automated and a CAQC system has been installed. The caster went into operation early 1988.

The CAQC system contains the full scope of the comprehensive CAQC package functions i.e.

- quality planning
- display of casting practice information (see fig. 13)
- process tracking
- quality prediction
- bloom disposition

The quality control system covers the production steps starting from the BOF
vessel via the ladle treatment stations to the continuous casting machine and
therefore collects data like
- BOF vessel data
- ladle treatment station data
- RH station data
- ladle furnace station data
- analysis data
- casting process data

As an example the data collected from the casting stage are listed in table 1.

The predicted bloom quality is determined in terms of the below listed quality
characteristics
. overlaps, teeming interruption marks
. oscillation marks
. pinholes, blowholes
. longitudinal lateral cracks
. longitudinal corner cracks
. transverse lateral cracks and transverse corner cracks
. nonmetallic macro inclusions, cross section
. nonmetallic macro inclusions, surface area
. internal cracks close to surface
. internal cracks
. pipe formation, central segregation
. slag clouds
. tension cracking

An example of a quality record is given in fig. 14.

The bloom disposition results in one of the following possible decisions
1 - quality o.k.
2 - inspection of surface
3 - light surface treatment (scarfing grinding)
4 - cold sampling
5 - cold sampling and inspection
6 - cold sampling and surface treatment

CAQC system for VOEST-ALPINE Donawitz works

The processing line of the VOEST-ALPINE Donawitz steelworks is illustrated in
fig. 15. The primary units are 1 KVA converter and 2 bottom-stirred LD converters.
Either the KVA and one LD converter or 2 LD converters are in operation
simultaneously. The KVA converter supplies the feed metal either to the LD
converter where, according to the quality requirements, 50 - 100% KVA metal are
input, or to the ladle furnace where 100% KVA metal are further processed. Ladle
metallurgy is done in the conditioning unit and the ladle furnace; an RH-degassing
station is presently under construction.

The total quantity of the so produced steel is cast on a 6-strand billet caster
(open-stream casting iwth and without gas protection as well as slide gate casting)
and on a 3-strand bloom caster (Pochmarski 1988). The casting sections are on
the billet caster 130 mm sq and on the bloom caster 225 mm sq. as well as
250 x 360mm (preferably input material for rails).

Regarding the introduction of CAQC figure 16 illustrates the quality variables
which are significant for the Donawitz product mix. Likewise, this figure gives
an overview of the testing method employed for the detection of the various kinds

of defects, as well as the criteria for assessment of these defects. Usually, standards which have been worked out in-house are used for reference, keeping expenditure to a minimum.

Figure 17 outlines maximum allowable rating values for quality variables at the cast product. In addition, the finished product and/or the minimum permissible reduction ratio are shown.

In the first implementation phase variant 3 of the above mentioned variants will be realized with the intent of extension to variant 4 in the future, so that the maximum allowable rating values will finally be graduated from 0 to 9, depending on the product requirements. The rating refers to various descriptions or standards and informs about the maximum allowable value of each quality variable, depending on the steel quality and requirements on the product to be supplied. The system is presently in the implementation phase so that operational experience cannot be given at present.

CAQC is a very useful tool for assuring high quality standards orientated towards final use of the product. Five variants of staggered complexity have been developed to suit the purpose and the available automation environment.

Economic advantages of the CAQC-system arise from substantial savings by an increase of yield, by the reduction of scarfing losses, and by an increase of hot charging. These positive effects result in particular from the complex evaluation of a great number of process parameters, so that unfavourable conditions can be rated with respect to the other parameters, which might compensate negative effects. In contrast, conventional practice leads to downgrading and conditioning if only a single process parameter is out of range.

In addition to these direct benefits CAQC results in continuous quality and productivity improvements.

REFERENCES

Fastner, T. and L. Pochmarski, (1982) Berg-Huttenmannische Monatshefte, 127, 327 - 333.

Fastner, T., A. Mayr, P. Narzt, and F. Wallner, (1988) Implementation of a computer aided quality control System for CC slab production at VOEST-ALPINE Linz. 4th Int. Conference Continuous Casting, Brussels Preprints No. 1, 142 - 154.

Krebs, H.G., H. Preissl, P. Schrack, and T. Fastner, (1987) Progress in automation of continuous casting machines. 4th VOEST-ALPINE Continuous Casting Conference, Linz, paper No. 22.

Pochmarski, L., A. Moser, W. Duetzler, and K. Schwaha, (1988) Production of high-quality long products at VOEST-ALPINE Donawitz works; McMaster Symposium Preprints, Hamilton, paper No. 8.

TABLE 1 Data tracked from casting operation at Posco bloom caster

a) continuous parameters
      net steel weight in tundish
      steel temperature in tundish
      oscillation frequency, actual
      casting speed
      temperature difference of prim. cooling water
      flow rate of gas for ladle shroud
      flow rate of gas for stopper
      type of mould powder
      secondary cooling program number
      free oxygen content in tundish
      immersion depth of snorkel
      tundish superheat
      first heat of a sequence or
      tundish fly
      change in casting speed
      oscillation stroke
      mould coating material
      number of heats cast with the mould
      alignment between mould and footroll
      alignment between footroll and bender
      alignment bender and casting-bow
      condition of mould plate

b) events
      position of ladle shroud
      ladle shroud off
      oxygen lancing of ladle slide gate
      tundish fly
      turret arm lifting
      lifting of tundish
      casting, start of the sequence
      casting, end of the sequence
      defect of snorkel tube
      cleaning of snorkel tube
      secondary cooling status
      mould level status
      snorkel tube status
      utilization EMS (y/n)
      mould level status
      mould powder change or deslagging

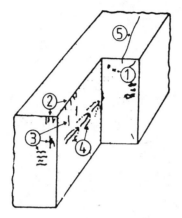

1. Macro Inclusions

2. Cracks Close to the Surface

3. Internal Cracks

4. Pipes, Center Segregation

5. Tension Cracks

Fig.1 Quality features of bloom/billet internal quality.

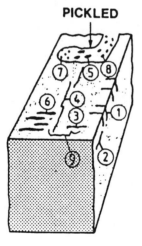

1. Transverse Corner Cracks

2. Longitudinal Corner Cracks

3. Transverse Mid Face Cracks

4. Longitudinal Mid Face Cracks

5. Slag Clouds

6. Oscillation Marks

7. Blowholes, Pinholes

8. Macro Inclusions

9. Overlaps

Fig.2 Quality features of bloom/billet surface & subsurface quality.

No.	Position	Quality Features (Quality Variables)	Quantification	
			Test method	Criteria/Unit
A	(sub)-surface slab	longit. wideface-	visual inspection / as cast	crack length/ slab length mm/m
B		longit. corner-	as cast	crack length/ slab length mm/m
C		transv. wideface-	scarfed max.2%	slab length No./m
D		transv.corner-	scarfed max.2%	slab length No./m
E		star-	scarfed max.2%	slab length No./m
F		pin-/blowholes		slab length No./m
G		macro-		No./m^2 surface
H		Ti-oxide-		No./m^2 surface
I		Al-oxide-	S-print	No. clouds/dm^2
P	internal, slab	S-segregation	S-print	factor VA-stand
Q		CMn-segregation	macro etch	factor VA-stand
R		halfway cracks	S-print	l/q-mm/mm
S		Al-oxide	S-print	No.clouds/dm^2
W	internal plate	lamination	u.s. -test	factor (%area, e.f.s.*)

*e.f.s. ... equivalent flow size

Fig.3 Quality variables for slab casting.

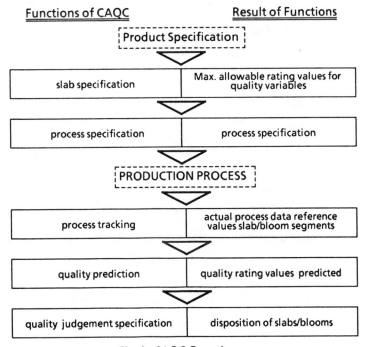

Fig.4 CAQC-Function

Step	Values for Process Parameters to be Detemined	Depending on
1	casting speed (Vg)	dimension analysis, rating values for G,H,I,P,Q,R
2	sec. cooling factor (s.c.)	Vg, analysis
3	S-content (caster related) (S_L)	Vg, analysis, rating values for C,P,Q,R
4	casting powder (c.p.)	Vg, analysis
5	steel temp. tundish °C above liquidus (t_T)	analysis, rating values for G,H,I,P,Q
6	bubbling amount, Ar (b.a.)	Vg, analysis, rating values for G,H,I,S
7	H-sample ladle	rating value for W

Fig.5   List of process parameters -slab casting.

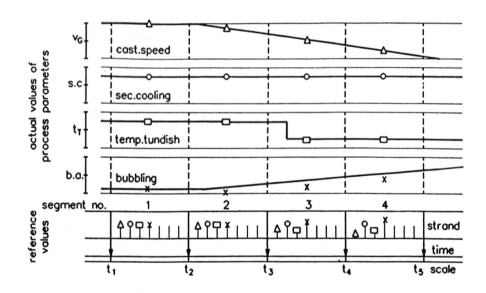

Fig.6   Process tracking (schematic)

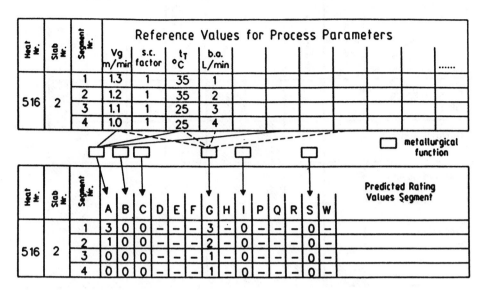

Heat Nr.	Slab Nr.	Segment Nr.	Reference Values for Process Parameters									
			Vg m/min	s.c. factor	t_T °C	b.o. L/min						......
516	2	1	1.3	1	35	1						
		2	1.2	1	35	2						
		3	1.1	1	25	3						
		4	1.0	1	25	4						

☐ metallurgical function

Heat Nr.	Slab Nr.	Segment Nr.	A	B	C	D	E	F	G	H	I	P	Q	R	S	W	Predicted Rating Values Segment
516	2	1	3	0	0	–	–	–	3	–	0	–	–	–	0	–	
		2	1	0	0	–	–	–	2	–	0	–	–	–	0	–	
		3	0	0	0	–	–	–	1	–	0	–	–	–	0	–	
		4	0	0	0	–	–	–	1	–	0	–	–	–	0	–	

Fig.7 Quality prediction (schematic)

heat	slab	steel grade	rating value	quality variable													
				A	B	C	D	E	F	G	H	I	P	Q	R	S	W
516	2	St 14-05	max. allowable	3	3	3	-	-	-	3	-	3	-	-	-	3	-
			predicted	3	0	0	-	-	-	3	-	0	-	-	-	0	-

slab judgement � specification

heat 516, slab 2 St 14-05

disposition		A	B	C	D	E	F	G	H	I	P	Q	R	S	W
	spot scarfing														
	test scarfing														
	machine scarfing														
	dawn grading														
	direct charging	x	x	x				x		x				x	

final decision : direct charging

Fig.8 Disposition of cast product: example for slab casting.

CAQC-Functions / CAQC-Variations	Quality planning	Dynamic determination of production parameters	Display of production parameters	Process tracking	Determination of deviations between planned and actual practice	Quality prediction	Disposition	Statistics and reporting
Defect Tracking (SPC)				▨				▨
Quality Tracking			▨	▨	▨			▨
Standard Quality Prediction						▨		▨
Advanced Quality Prediction						▨	▨	▨
Comprehensive Quality Control	▨	▨	▨	▨		▨	▨	▨

Fig. 9 Computer aided quality control overview CAQC-variations.

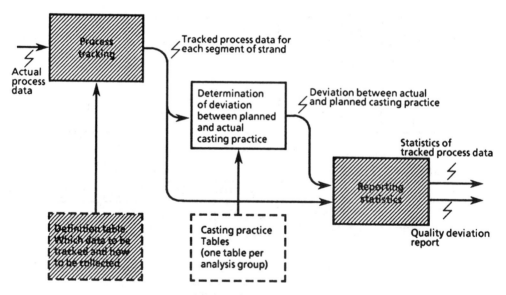

Fig.10 Computer aided quality control
Variation 1: Defect tracking (hatched blocks only)
Variation 2: Quality tracking (all blocks)

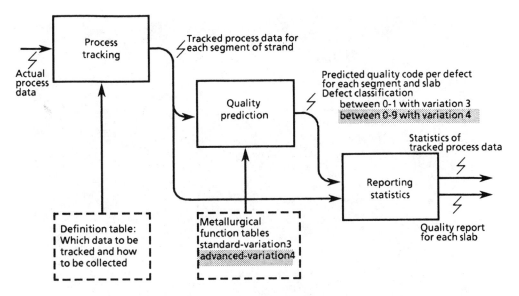

Fig.11 Computer aided quality control -Variation 3: Standard quality prediction.
-Variation 4: Advanced quality prediction
(including dotted blocks).

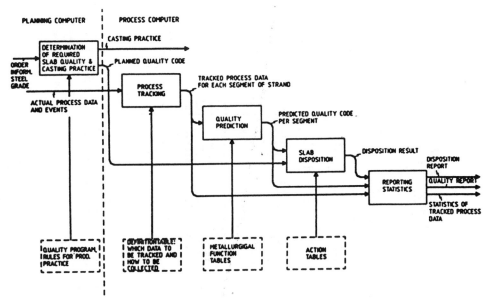

Fig. 12 Computer aided quality control -Variation 5: Comprehensive quality control.

SHIFT: A   BLOOM CASTER   88-05-25 13:10:10

CASTER PRACTICE MONITOR

STEEL GRADE   SUP9A

SHROUD	gas type	AR	flow rate of gas	10 [1/min]
TUNDISH	lining material	N	type of dam and/or weir	A
	aim temperature	1503 [C]	liquidus temperature	1478 [C]
SNORKEL	type	S	flow rate of gas	3 [1/min]
MOULD	coating material	O	taper	0.8 [%]
PRIMARY COOLING	temperature difference	9.0 [C]		
CASTING SPEED	minimum	70 [cm/min]	maximum	80 [cm/min]
SECONDARY COOLING	number	1 mild		

Fig.13 CAQC at Posco bloom caster (example of production practice).

Shift: A   BLOOM CASTER   89-05-10 16:00:-1

BLOOM REPORT

Heat: X00935

MARKING NUMBER	PLANNED LENGTH [mm]	ACTUAL LENGTH [mm]	ACTUAL WEIGHT [ton]	PLANNED QUALITY CODE	ACTUAL QUALITY CODE	VISUAL CODE	DISPO CODE	TIME OF CUT	SAMPLE LENGTH [mm]
X009351010	6990	7090	4.1	333333333333	0220003100000		0	15:02:10	:
X009351020	6990	7090	4.1	333333333333	0220003100000		0	15:11:16	:
X009351033	6990	7090	4.1	333333333333	0440005700000		3	15:19:56	:
X009351040	6990	7090	4.1	333333333333	0330001100000		0	15:28:05	:
X009351050	6990	7090	4.1	333333333333	0220001100000		0	15:37:31	:
X009351060	6990	7090	4.1	333333333333	0220001100000		0	15:46:51	:
STRAND 1 SUM:	41940	42540	24.5						
X009352010	6990	7090	4.1	333333333333	0220003100000		0	15:00:30	:
X009352020	6990	7090	4.1	333333333333	0220003100000		0	15:09:27	:
X009352033	6990	7090	4.1	333333333333	0550003700000		3	15:18:22	3
X009352040	6990	7090	4.1	333333333333	0220003100000		0	15:27:06	:
X009352050	6990	7090	4.1	333333333333	0220001100000		0	15:36:11	:
X009352060	6990	7090	4.1	333333333333	0220001100000		0	15:45:23	:
STRAND 2 SUM:	41940	42540	24.5						
X009353010	6990	7090	4.1	333333333333	0220003100003		0	15:00:59	:
X009353020	6990	7090	4.1	333333333333	0220003100000		0	15:10:13	:
X009353035	6990	7090	4.1	333333333333	7330005700000		5	15:18:50	:
X009353042	6990	7090	4.1	333333333333	4220001100000	73	2	15:28:01	:
X009353050	6990	7090	4.1	333333333333	0220001100000		0	15:37:24	:
X009353060	6990	7090	4.1	333333333333	0220001100000		0	15:46:52	:
STRAND 3 SUM:	41940	42540	24.5						
X009354010	6990	7090	4.1	333333333333	0220003100000		0	15:02:47	:
X009354020	6990	7090	4.1	333333333333	0220003100000		0	15:11:09	:
X009354030	6990	7090	4.1	333333333333	0220001100000		0	15:20:44	:
X009354040	6990	7090	4.1	333333333333	0220001100000		0	15:29:05	:
X009354053	6990	7090	4.1	333333333333	0220001100000		0	15:38:37	:
X009354060	6990	7090	4.1	333333333333	0220001100000		0	15:47:59	:
STRAND 4 SUM:	41940	42540	24.5						
TOTAL SUM:	167760	170160	98.1						

Fig. 14 CAQC at Posco bloom caster (example of quality report).

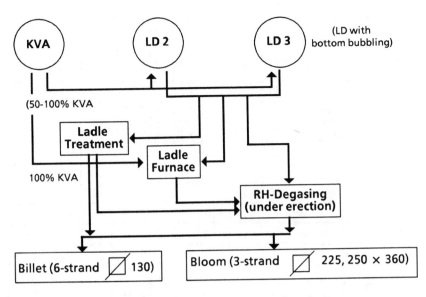

Fig. 15 Process route in Donawitz-plant.

position	No.	quality variables	quantification test method	criteria/unit
(sub) surface	A	overlaps, teeming interruption marks	visual inspection as cast	N/m
	B	oscillation marks	visual inspection as cast	standard
	C	pinholes/blowholes	visual inspection scarfed or ground	N/m^2
	D	longitudinal lateral cracks	visual insp. sand blasted	standard
	E	longitudinal corner cracks long. cracks close to corner	visual inspection sand blasted	standard
	F	transverse lateral cracks transverse corner cracks	visual inspection sand blasted	standard
	G	nonmetallic macro inclusions (surface area)	visual inspection scarfed or ground	N/m^2
	H	slag clouds (nests)	visual inspection as cast	N/m^2
internal	J	nonmetallic macro inclusions (cross section)	macro-etch transversal	N/dm^2
	K	internal cracks close to surface	macro-etch transversal	standard
	L	internal cracks	macro-etch transversal	standard
	M	pipe formation central segregation	macro-etch longitudinal	standard
surface-internal	N	tension cracks	visual inspection as cast macro-etch transversal	occurance

Fig. 16 Quality variable bloom/billet- example for Voest-Alpine Donawitz works

steel grade	final product (deformation ratio)	max. allowable rating values (RV) of quality variables												
		A	B	C	D	E	F	G	H	J	K	L	M	N
concrete steel	> 4 × bar, wire	2	8	5	3	3	3	4	4	9	7	8	9	no occurance
structural steel	> 8 ×	1	6	3	2	2	2	2	2	6	5	6	7	no occurance
extra low carbon steel	wire rod	0	3	1	1	1	1	1	1	2	3	4	5	no occurance
medium alloyed steel	> 12 ×	0	3	0	0	0	0	0	0	0	2	2	4	no
high carbon steel	wire rod	0	5	1	1	1	1	1	1	2	3	4	3	no
special high														
carbon steel	wire rod	0	4	0	0	0	0	0	0	1	2	4	3	no
spring steel	> 12 × flat, wire	0	3	0	0	0	0	1	0	1	0	3	4	no
free cutting steel	> 15 × bar, wire	0	5	3	3	3	3	3	1	3	4	4	6	no occurance
bearing steel	> 25 ×	0	3	0	0	0	0	0	0	0	2	2	3	no
rail steel	≥ 10 ×	0	3	0	0	0	0	0	0	1	0	2	3	no
seamless tube steel	seamless tubes (max∅167)	0	3	0	0	0	0	0	0	2	0	3	5	no

Fig. 17   Example of maximum allowable rating values for product mix at Voest-Alpine Donawitz.

LASER-ULTRASONICS: A NOVEL SENSOR TECHNOLOGY FOR THE STEEL INDUSTRY

J.-P. Monchalin*, J.-D. Aussel*, R. Héon*, J.F. Bussière*
P. Bouchard** and J. Guévremont**
* National Research Council Canada
Industrial Materials Research Institute
75 De Mortagne Blvd.
Boucherville, Québec, Canada   J4B 6Y4
** Tecrad Inc
1000 Ave. St. Jean Baptiste
Québec, Québec, Canada   G2E 5G5

ABSTRACT

Laser-ultrasonics is a novel technology which uses lasers to generate and detect
ultrasound at a distance and permits to eliminate the problem of ultrasonic
coupling to hot products. The principles and the implementation of this technology
are presented. This technology is being considered to be applied to on-line
measurement of the wall thickness of seamless pipes, to the detection of surface
and buried defects in steel products and to grain size evaluation. Perspectives
and on-going projects in these areas are reviewed.

KEYWORDS

Ultrasonics, ultrasound, laser-ultrasonics, nondestructive testing, industrial
sensors, sensors for steel industry, process control, ultrasonic attenuation, grain
size measurement.

INTRODUCTION

The steel industry, like many other industries, will benefit from a better control
of its manufacturing processes and of the quality of its products. Proper control
of any process requires an appropriate sensor to measure a physical parameter on
the product during processing, such as shape parameters, temperature, microstructure characteristics ... It is also important to insure that the product is free
of inner or surface flaws or at least does not have flaws exceeding a prescribed
size or distribution. Various process and quality control sensors are being used
in the steel industry and many others are in development, based on optics, thermo-
graphy, X-rays, magnetism, ultrasonics ...

Ultrasonics, in particular, is known to be useful for measuring thicknesses,
detecting internal and surface defects and for providing information on microstruc-
tural parameters such as grain size. Ultrasonics require a medium for coupling
ultrasound from the transducer to the workpiece and this is provided in the
classical implementation of the technique either by direct contact using a thin
fluid layer, by a water bath or by a water jet. Obviously, when the inspected part
is at elevated temperature, such fluid coupling is prohibited and a noncontact
method is required. Two methods which permit to generate and detect ultrasound
without contact are known, one is based on electromagnetic-acoustic transducers
(EMATS) and the other one is laser-ultrasonics.

EMATS, which consist of an electrically excited coil located within the field produced by a magnet generate ultrasound by the Lorentz force on the induced eddy currents or by magnetrostriction (Frost, 1979). The inverse principle is used for detection. Since their sensitivity depends on the magnetic properties of steel, their performance is temperature dependent, but their main limitation is the requirement to be relatively close to the surface (mm range). This means that the device has to be moved in and out from proximity to a very hot product and that the distance to the surface should be maintained as constant as possible. An EMAT system has been experimented on a continuous caster (Kawashima, 1983) and on a seamless pipe production line (Yamaguchi, 1985) and development of this technology is continuing in several countries (Böttger, 1987, Burns, 1988).

Laser-ultrasonics, which uses lasers to generate and detect ultrasound is on the other hand free from any proximity requirement and can operate meters away from the inspected product. We present below the principles of the laser-ultrasonic technique, including the generation and detection aspects, focusing the presentation mostly on the principles applicable to on-line inspection in the steel industry. We then review the latest developments and the prospects of application of this emerging technology to thickness gauging at elevated temperature, detection of inner and surface defects and microstructure determination.

PRINCIPLE OF LASER-ULTRASONICS

Laser-ultrasonics uses a short pulse high power laser for ultrasound generation and another laser which is pulsed or continuous for the detection of the ultrasonic surface motion, as sketched in Fig. 1. Generation and detection can be made from opposite sides of the workpiece or from the same side and even at the same location.

GENERATION OF ULTRASOUND WITH LASERS

Two methods can be used (Scruby, 1982; Hutchins, 1986). At low laser power density, there is no phase change at the surface and only transient surface heating which produces essentially tangential stresses (thermoelastic regime). At higher laser power density there is surface melting and surface vaporization giving near the surface a hot expanding plasma (ablation regime). The ultrasonic stress in this case is produced by the recoil effect following material ejection and is essentially normal. A longitudinal wave is then emitted normally to the surface. Ultrasonic stresses in this case are comparable in magnitude to the ones produced by conventional piezoelectric transducers using peak voltage excitation of a few hundred volts, whereas they are weaker in the thermoelastic regime. Therefore, for probing hot steel specimens, the ablation regime will be preferred. Ablation causes the removal of the surface oxide and has no adverse or durable effect on the steel product. For the detection aspect, it is useful to know that the laser spot size determines the beam spreading within the specimen (diffraction effect).

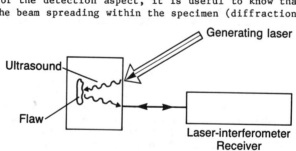

Fig. 1:   Schematic of principle of the laser-ultrasonic technique.

Surface waves are also emitted by the laser impact. Their wavelength and spectral contents are generally determined by the size of the impact and not by the laser pulse duration. Using cylindrical optics, a line can be projected on the surface, which gives a linear surface wave wavefront. By using conical optics (axicon), a circle can be projected giving a converging surface wave (Cielo, 1985). At the point of convergence, there is a strong ultrasonic deformation which can be more easily detected.

Concerning the technological aspect, the commercially available lasers suitable for generation by ablation and with repetition rates exceeding 20 Hz are Neodymium-YAG lasers and excimer lasers. Nd-YAG lasers operate at 1.06 μm (10 ns pulse typical, up to 1 J per pulse) and can be doubled (532 nm, 300 mJ typical) and tripled (355 nm, 150 mJ typical). Excimer lasers operate at higher repetition rates (>50 Hz) in the ultraviolet (XeF: 351 nm, XeCl: 308 nm, KrF: 248 nm) with energies up to 1 J. Since high peak laser power is used, either in the visible or the ultraviolet, transmission through an optical fiber, even with a large core, is limited by material breakdown and does not seem feasible at this time. Therefore, the generating laser beam should be either directly coupled onto the specimen or steered upon it with mirrors, which requires a suitable enclosure to protect the laser and its optics from the harsh environment generally encountered in the steel industry.

OPTICAL DETECTION OF ULTRASOUND

The various methods of detection of ultrasound at the surface of opaque solids have been recently reviewed (Monchalin, 1986). A method can be considered suitable for hot steel inspection if it permits to collect scattered light from a rough surface over a sufficiently large spot and if it is very insensitive to low frequencies (<1 MHz). This last requirement is particularly important when detection is performed at the same location as generation since the generated plasma causes a very strong pertubation on the optical wave. In order to understand the various methods and their detection principles, it is useful to know that the effect of ultrasound on the scattered light can be described either as a phase modulation or a Doppler shift of the instantaneous frequency or the generation of optical sidebands (see Fig. 2). Optical heterodyning which consists in making the wave scattered by the surface to interfere with a reference wave coming directly from the laser (i.e. single frequency, without sideband) would be usable when the scattered wave is sent onto a phase-conjugating mirror (Paul, 1987), which makes the light to retrace its path back upon itself and avoids the speckle problem associated to surface roughness. Another technique, based on the same principle, which derives its reference wave from the wave scattered by the surface by removing the sidebands (Monchalin, 1988) could be used as well. Both these techniques permit broadband detection, which is particularly useful for some applications (see below). However, they do not appear at this time well developed and do not seem to have been applied yet to hot steel probing.

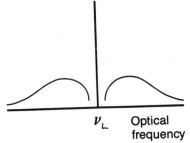

Fig. 2: Optical spectrum of the light scattered by ultrasound following pulsed laser excitation. ν_L is the laser frequency.

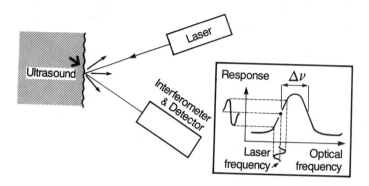

Fig. 3:   Principle of detection by optical frequency demodulation
          using a velocity or time-delay interferometer.

The other interferometric detection method, which is not only suitable to hot steel
inspection but also has been used as well, is called velocity or time-delay inter-
ferometry.   In this method, there is no physically distinct reference wave as in
optical heterodyning and there is only the scattered wave which is sent onto the
interferometer.   The interferometer works essentially as a light filter and demodu-
lates the frequency modulated light as sketched in Fig. 3.

Two types of interferometers can be used, either two-wave interferometer (Michelson
or Mach-Zehnder) or a multiple-wave interferometer (Fabry-Pérot).   The two-wave
approach was initially taken by Krautkrämer in Germany (Kaule, 1976) and is being
pursued by the Betriebsforschunginstitut (BFI) in Düsseldorf, Germany (Keck,
1987).   The use of this interferometer is illustrated in Fig. 4.   The frequency
response of such a

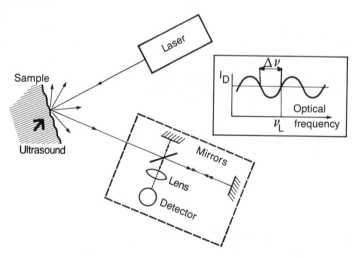

Fig. 4:   Detection of ultrasound with a two-wave (Michelson) velocity interfero-
          meter. $\Delta \nu = 1/(2\tau)$ τ being the propagation delay time between the two
          interferometer arms.

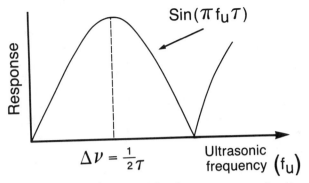

Fig. 5:   Frequency response (magnitude) of a two-wave velocity interferometer.
The phase spectrum is a linear function of frequency.

receiver is not uniform and is sketched in Fig. 5. The system is very insensitive
at low frequencies and its maximum sensitivity occurs when the ultrasonic frequency
is equal to $\Delta \nu = 1/(2\tau)$. In order to have adequate sensitivity in the MHz range,
the path difference between the two arms should be several meters, which makes such
a system large and bulky. Adequate light gathering capability from a rough surface
required the addition in one of the paths of a liquid column as described by
Monchalin (1986).

Reduced interferometer size is obtained by using a multiple-wave interferometer
(Fabry-Pérot), which easily permits by appropriate choice of mirror reflectivity to
obtain a sufficiently narrow bandwidth. When the two mirrors are concave and
confocal (which means that the center of either mirror is found on the other) a
large light gathering efficiency is obtained. Also, with this interferometric
system, unlike the classical two-wave interferometers, mirror orientation is unim-
portant. The principle of operation of such an interferometer is explained by
Fig. 6, which also shows the optical frequency response. A typical ultrasonic
frequency response is shown in Fig. 7. Such an approach has been taken by the
Industrial Materials Research Institute (IMRI) of the National Research Council of
Canada (NRCC) (with the assistance of CANMET of Energy, Mines and Resources Canada)
and is being further developped by Tecrad Inc. in collaboration with IMRI/NRCC.
The system under development and scheduled for in-plant trial in a near future is
sketched in Fig. 8. The system is made up of two units linked by optical fibers.
The generating unit includes the generating laser which is presently a doubled
Q-switched Nd-YAG laser and the beams mixing and light collecting optics (15 cm in
diameter). This unit is designed to generate and detect at 1.5 m and can operate
at shorter or longer distances by changing the light collecting optics. The
receiving unit includes the receiving Nd-YAG laser (50 Hz, 50-100 μs pulses,
several KW peak power) and the confocal Fabry-Pérot receiver. This unit is
followed by digital sampling and digital processing electronics.

After having presented the principles of generation and detection and their implementation, we describe below potential applications in the steel industry, namely
thickness measurement, defect detection and microstructure monitoring.

THICKNESS GAUGING

Ultrasonics is widely used for thickness gauging on steel and other materials.
Since the technique measures only the propagation time of ultrasound through the
sample thickness, the determination of this thickness requires the knowledge of the
propagation velocity in the material. This can be obtained by calibration on a

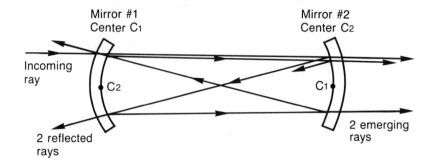

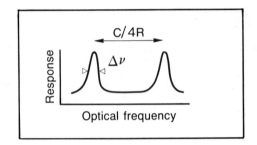

Fig. 6: Confocal Fabry-Pérot interferometer. The insert shows the optical frequency response. R is the radius of curvature of the mirrors and C is the speed of light.

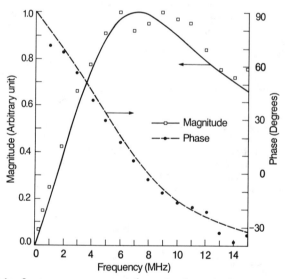

Fig. 7: Ultrasonic frequency response (magnitude and phase, theoretical and experimental) of a Fabry-Pérot receiver (R = 50 cm, $\Delta\nu$ = 10 MHz).

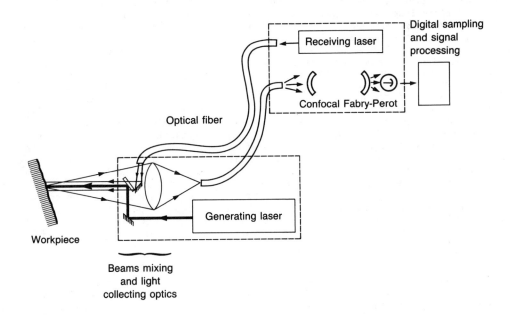

Fig. 8:   System in development.

specimen of known thickness. The same method can be applied on hot or fast moving
samples, such as strips, using the laser-ultrasonics technique. Laser-ultrasonics
has several clear advantages over X-ray gauges because it is a truly remote technique and easy scanning across a strip can be performed by reflecting the beams off a
rotating mirror. However, the present accuracy of X-ray gauges ($\approx \pm 0.1\%$) (Leaver,
1984) may be difficult to match, essentially because of ultrasonic velocity changes
by possible texture variations during rolling and across the strip. However, this
application does not seem to have been sufficiently studied to rule out completely
the use of laser-ultrasonics.

A thickness measurement which appears more promising, because in this case X-ray
gauges cannot be used, is the measurement of the wall thickness of pipes,
especially seamless pipes. This application has been recently explored in a mill
environment by BFI (Keck, 1987). All the laser interferometric and electronic
equipment was located in a kind of mobile laboratory using a container of large
size. The measurements were made at the output of the first piercing machine on
tubes at 1230°C located 5 m away from the front lens of the system which was 30 cm
in diameter. The generating laser used was a KrF excimer laser (0.7 J, 15 ns per
pulse) which also had the effect of removing the oxide layer on the surface. Oxide
removal gives a nonabsorbing surface of much higher reflectivity which eases detection. BFI results have shown a signal-to-noise ratio of typically 2 or 3, sometimes 1, but nevertheless have permitted to show thickness variations along hot
pipes. This work demonstrates the potential of the technology for controlling the
hot piercing process.

At IMRI, a similar application is also now being pursued in association with Tecrad
Inc. and Algoma Steel Inc., a canadian seamless pipe manufacturer. The

experimental setup is the one described above and sketched in Fig. 8. Figure 9 shows the first laser shot signal observed at 916°C on a specimen cut from a seamless pipe 14.3 mm thick and 24.5 cm in diameter. The specimen was heated in air for about 20 minutes in a preheated oven. Generating and receiving spot sizes were 5 mm in diameter. When samples are left in the oven for extended periods of time, much worse signal-to-noise conditions are generally observed. The causes of the decrease of the signal-to-noise ratio have not been clearly identified at this time. They could include grain growth, which increases ultrasonic attenuation, additional noise of the denser and longer lasting plasma produced by thicker oxide coverage, possible additional ultrasonic noise at the source, less efficient coupling of ultrasound by semi-adherent and not completely removed oxide layers (nonadherent layers are scraped off), possible additional attenuation and pulse lengthening by oxide layers on the inner surface. The conditions prevailing on the hot worked pipe in the mill are likely to be better than the worst conditions encountered in the laboratory after extensive heating, but the signal-to-noise ratio which will be encountered in the mill can hardly be predicted at this time from laboratory experiments. Anyhow, we have also shown that, even in very poor signal-to-noise ratio conditions where the first echo cannot be distinguished from noise, digital sampling followed by cross-correlation permits to measure the time delay with reasonable precision. For example, for a signal-to-noise ratio of 1 between 3 to 5 MHz and a signal window of 2 μs, the expected precision is 25 ns or 0.4% for 15 mm thickness (Aussel, 1988).

As mentioned above, being able to measure the ultrasonic propagation time is not sufficient and the propagation velocity should be known. This velocity depends upon temperature and texture. The precision required for the temperature determination depends upon the temperature range of the ultrasonic measurement. As shown in Fig. 10, which shows the longitudinal velocity variation with temperature for a plain carbon steel (Monchalin, 1988b), the requirement is more severe in the transition range of ferrite to austenite. An accuracy of $\pm$ 0.1% requires a knowledge of the average through-thickness temperature within $\pm$ 8°C in the austenitic range and within $\pm$ 2°C at the steepest slope point of the transition zone. Since temperature is measured on the surface, a sufficiently accurate model is also needed to predict bulk temperature. The effect of texture variations along and around the pipe and from pipe to pipe need also to be evaluated.

FLAW DETECTION

Laser-ultrasonics, like any ultrasonic technique, could be used to detect flaws at the surface or in the bulk of a material. The technique has the additional advantage of not being limited by temperature. Little work in conditions similar to those encountered in industry and on industrial defects (such as the defects of cast slabs) seems to have been done so far. We show in Fig. 11 the detection of a side-drilled hole in a hot rolled plate (black oxidized surface) at room temperature which was reported earlier (Monchalin, 1988b). The use of laser generated surface waves has also been explored by various researchers on polished surfaces using either line generation (Aindow, 1982) or circle generation (Cielo, 1986).

GRAIN SIZE EVALUATION

It is well known that grain size affects the velocity and the attenuation of ultrasound propagating in steel (Stanke, 1984). The effect upon velocity is small, whereas the one on attenuation could be very large if sufficiently high ultrasonic frequencies are used. In spite of that, the evaluation of grain size from ultrasonic attenuation is generally difficult because the apparent attenuation caused by beam spreading (diffraction effect) has to be evaluated. This is further aggravated in the case of laser-ultrasonics by the fact that laser beams rarely have a well defined intensity distribution. We have shown that the generation and recep-

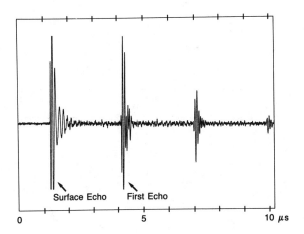

Fig. 9: Echos observed on a 14.3 mm thick specimen cut
from a seamless pipe and heated to 916°C

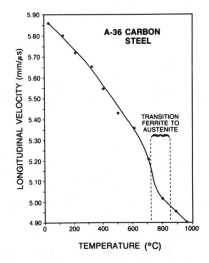

Fig. 10: Variation of longitudinal ultrasonic velocity of A-36 carbon
steel with temperature (Monchalin, 1988).

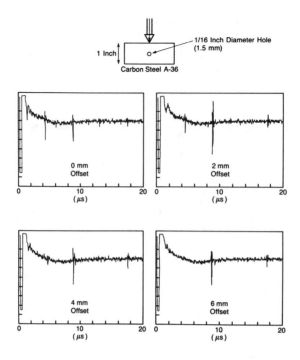

Fig. 11:   Detection of a side-drilled hole by laser-ultrasonics on a hot-rolled plate at room temperature. A Q-switch Nd-YAG laser was used for generation and a one-watt CW argon ion laser for detection. The offset of the generating/receiving location is measured from the vertical passing through the hole center (Monchalin, 1988b).

tion spots should be sufficiently large and of about equal size, in order to make the diffraction correction negligible (Aussel, 1988b). Figure 12 shows the first ever reported to our knowledge ultrasonic attenuation spectrum of a carbon steel plate, obtained by laser-ultrasonics. The system shown in Fig. 8 was used for the measurement. The data follows the expected f^4 variation (f is the ultrasonic frequency) corresponding to Rayleigh scattering, i.e. to the case where the ultrasonic wavelength is much larger than the grain size (for this specimen the grain size is $\approx$ 14 μm). Figure 12 shows that there is good agreement with measurements made with conventional piezoelectric transducers. Also shown in Fig. 12 are the results obtained with spots of very different sizes. In this case, the diffraction correction becomes very large and since it cannot be precisely evaluated because of lack of knowledge of the laser beam intensity distributions, the measurement is not possible. This work will be extended to smaller grain sizes and to elevated temperature conditions in the near future.

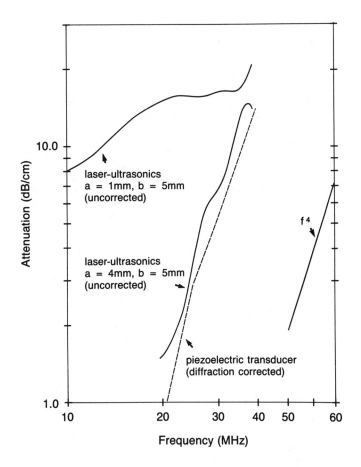

Fig. 12: Longitudinal ultrasonic attenuation of a hot-rolled steel plate measured by the laser-ultrasonic system shown in Fig. 8. a and b are the radii of the generation and reception spots, respectively (Aussel, 1988).

SUMMARY AND CONCLUSION

We have presented the methods of ultrasound generation and detection by lasers suitable for application in the steel industry. We have described recent developments to apply them to the measurement of the wall thickness of hot seamless pipes, to defect detection and grain size evaluation. The seamless pipe application is the one which has been so far the object of the most advanced experimentation, including mill tests and which presents good prospects to be actually used in a mill. Flaw detection has been mostly evaluated at the principle level and extensive work is needed on practical flaws before the potential of application of the laser-ultrasonic technology to flaw detection is assessed. In the case of grain size evaluation, a method was found which permits to avoid the difficulties caused by diffraction and unknown laser beam distributions, so the perspectives of practical use appear excellent.

REFERENCES

Aindow, A.M., Dewhurst, R.J. and Palmer, S.B. (1982). Laser-generation of directional surface acoustic wave pulses in metals, Optics Com., vol. 42, pp. 116-120.

Aussel, J.-D. and Monchalin, J.-P. (1988a). Precise ultrasonic velocity and acoustic constants determination by laser-ultrasonics, unpublished.

Aussel, J.-D. and Monchalin, J.-P. (1988b). Measurement of ultrasound attenuation by laser-ultrasonics, unpublished.

Böttger, W., Graff, A., Schneider, H. (1987). Dickenmessung an Stahl mit electromagnetischer Ultraschallanregung bei Temperatur bis 1200°C, Materialprüfung, vol. 29, pp. 124-128.

Burns, L.R., Alers, G.A., MacLauchlan, D.T. (1988). A compact EMAT receiver for ultrasonic testing at elevated temperatures, in Review of Progress in Quantitative NDE, vol. 7B, Plenum Press, N.Y. pp. 1677-1683.

Cielo, P., Nadeau, F. and Lamontagne, M. (1985). Laser generation of convergent acoustic waves for materials inspection, Ultrasonics, vol. 23, pp. 55-62.

Cielo, P., Jen, C.K., Maldague, X. (1986). The converging surface acoustic wave technique: analysis and application to nondestructive evaluation, Can. J. Physics, vol. 64, pp. 1324-1329.

Frost, H.M. (1979). Electromagnetic-Ultrasound transducers: principles, practice and applications, in Physical Acoustics (W.P. Mason, ed.), vol. 14, Academic Press, N.Y., pp. 179-275.

Hutchins, D.A. (1986). Mechanisms of pulsed photoacoustic generation, Can. J. Physics, vol. 64, pp. 1247-1264.

Kaule, W. (1976). Laser induced ultrasonic pulse testing, in Proceedings of the 8th World Conference on Nondestructive Testing, paper 3J5.

Kawashima, K., Soga, M. and Iwai, K. (1983). Electromagnetic ultrasonic inspection and its applications. Nippon Steel Technical Report, No. 21, pp. 315-329.

Keck, R., Krüger, B., Coen, G. and Häsing, W. (1987). Wanddickenmessung an 1230°C heissen Rohrluppen mit einem neuartigen Laser-Ultraschall-System, Stahl und Eisen, vol. 107, pp. 1057-1060.

Leaver, E.W. (1984). Thickness and thickness profile gauges for hot strip in steel rolling mills. Report to IMRI, unpublished.

Monchalin, J.-P. (1986). Optical detection of ultrasound, IEEE Trans. on Ultrasonics, Ferr. and Frequency Control, vol. UFFC-33, pp. 485-499.

Monchalin, J.-P. (1988a). Broadband optical detection of transient motion from a scattering surface, unpublished.

Monchalin, J.-P., Aussel, J.-D., Bouchard, P., and Héon, R. (1988b). Laser-ultrasonics for industrial applications, in Review of Progress in Quantitative NDE, vol. 7B, ed. by D.O. Thompson and D.E. Chimenti, Plenum Press, N.Y., pp. 1607-1614.

Paul, M., Betz, B. and Arnold, W., (1987). Interferometric detection of ultrasound at rough surfaces using optical phase conjugation, Appl. Phys. Lett., vol. 50, pp. 1569-1571.

Scruby, C.B., Dewhurst, R.J., Hutchins, D.A., and Palmer, S.B. (1982). Laser generation of ultrasound in metals, in Research Techniques in Nondestructive Testing, vol. 2, R.S. Sharp editor, Academic Press, N.Y., pp. 281-327.

Stanke, F.E. and Kino, G.S. (1984). A unified theory for elastic wave propagation in polycrystalline materials, J. Acoust. Soc. Am., vol. 75, pp. 665-681.

Yamaguchi, H., Fujisawa, K., Murayama, R., Hashimoto, K., Nakanishi, R., Kato, A., Ishikawa, H., Kadowaki, T. and Sato, I. (1985). Development of a wall thickness measuring system using EMAT for hot seamless steel tubes and pipes, Proceedings of the 11th World Conference on NDT, Las Vegas, U.S.A., pp. 734-740.

ULTRASONIC MEASUREMENTS IN HOT AS-CAST BILLETS:
FIELD TESTING OF AN EMAT - EMAT APPROACH TO SENSING INTERNAL TEMPERATURES

J. R. Cook*, B. E. Droney**, and J. F. Jackson*

* Armco Inc., Armco Research & Technology, Middletown, OH
** Bethlehem Steel Corporation, Bethlehem, PA

ABSTRACT

This paper will comment on the field testing of a prototype sensing system for
monitoring internal temperatures in solid and solidifying materials. The system
was developed as part of an ongoing collaboration among AISI member companies, the
U. S. National Bureau of Standards and Battelle's Pacific Northwest Laboratories.
The approach uses "non-contacting" electromagnetic acoustic transducers (EMATs) in
a transmitter and receiver combination to monitor ultrasonic times-of-flight
(TOF). That data can be related to the internal temperature distributions in as-
cast materials. The field testing demonstrated the viability of the EMAT approach: the device survived difficult mill conditions and successfully acquired
data from moving, as-received strand cast materials with surface temperatures in
excess of 1130 Celsius.

KEYWORDS

Hot connection, direct rolling, direct connection, internal temperature, sensors,
EMAT, EMpulser, automatic control, process control, ultrasonic, field testing,
prototype, solidifying, non-contacting, as-cast.

INTRODUCTION:

Our American Iron and Steel Institute (AISI) Collaborative Task Unit (CTU) has
just completed a demonstration of the EMAT-EMAT approach to ultrasonic monitoring
of hot as-cast billets in a steel mill environment. This demonstration was the
concluding phase of a multifaceted effort to develop the technological basis to
meet a critical need: sensing the internal temperature distribution in solid and
solidifying bodies of steel.

The effort has been truly collaborative - involving active participation by eight
different steel companies (1), the U. S. National Bureau of Standards (NBS), and
Battelle's Pacific Northwest Laboratory (PNL). The work has been supported by
AISI member contributions, the U.S. Department of Energy and the U.S. Department
of Commerce.

HOT CONNECTION: SENSOR REQUIREMENTS

One of the steel industry's overriding concerns is to improve the coordination of processing and scheduling to take full advantage of the known yield, quality, productivity and energy advantages of hot connecting and direct rolling strand cast material.

Specific sensors which would aid materially in accomplishing these goals have been detailed (2). Ideally, these sensors would survive mill environments and function in real-time on as-produced materials receiving no special surface preparation. A priori knowledge of material properties, processing history and any conceivable ancillary measurements of surface temperatures, dimensions, etc. might be incorporated. The sensors have to provide reliable information in a cost-effective manner, and induce no unusual hazards to processed materials or personnel. Those sensors would:

- enhance control of strand casting processes to ensure metallurgical quality and maximize residual energy content,

- determine the workpiece temperature distribution before either direct rolling or reheating,

- monitor workpiece temperature profiles during reheating to optimize furnace heating and pacing conditions,

- estimate workpiece profiles for dynamic optimization of rolling and finishing processes.

PROCESS CONTROL: THE SENSOR GAP

The AISI had developed a comprehensive list of sensors which would substantially improve processing efficiencies (3). Some fell into a class for which no known sensing approach existed. Our AISI collaborative task unit (CTU 5-4) was organized to promote the development of a sensing approach which could meet some of the sensing requirements for hot connection: the determination of the temperature distribution in hot solid and solidifying materials. The task unit leveraged AISI member financial contributions, helped coordinate the technical efforts, and is now transferring this technology to the industry.

It was understood from the beginning that this would be a high risk undertaking. The profit motive was not in the sale of sensor systems or components, but in the use of the sensor system to monitor and control processes more precisely. The sole intent was to develop and demonstrate candidate technologies which instrument manufacturers or system integrators might be willing to exploit and bring to full commercialization.

INTERNAL TEMPERATURE SENSING: THE METHOD EVOLVES

Both eddy current and ultrasonic methods have been shown effective in interrogating internal temperature distributions. Each method relies on material parameters that have a definable and reproducible temperature dependence and each requires a mathematical reconstruction to extract the apparent temperature distributions. Sensitivities can be influenced by numerous material parameters (e.g. grain size, texture, strain, porosity, etc.) and these influences will prove useful in other sensor applications.

The eddy current approach depends strongly on electrical conductivity and geometric shape and is well suited to aluminum. The approach may also prove workable on steel products above the Curie temperature.

Ultrasonic approaches rely mainly on the measurable dependence of ultrasonic velocity with temperature (Fig. 1). The velocity can be determined from the ratio of two measured variables: the ultrasonic path length and time-of-flight (TOF). Each must be monitored accurately and simultaneously or at the same location under identical conditions.

The collaborative effort progressed through distinct phases with intensive developmental efforts at NBS and PNL, hardware integration and testing at ARMCO Research, and the very recent field testing on a commercial horizontal continuous caster at Baltimore Specialty Steel Corporation (BSSC).

a. "Proof of Concept" and Beyond (NBS)

The scientific basis for the ultrasonic approach was developed under the auspices of the Research Associate Program at NBS (4,5). That work demonstrated the "proof of concept" and emphasized:

. studies of ultrasonic sensors including pulsed lasers, conventional piezoelectric transducers, and various EMAT designs,

. the generation of reference data on the variation of ultrasonic velocity with temperature in different grades of steel,

. the laboratory measurement of ultrasonic TOFs in reheated samples of cylindrical and rectangular bodies of steel to 800 Celsius,

. the development of algorithms to reconstruct temperature distributions from such information,

. the integration of these components into a laboratory test fixture capable of determining and displaying internal temperature profiles in near real-time.

NBS continues to pursue collateral technologies in signal processing and error analyses and some uniquely innovative surface wave and dimensional resonance approaches. They are currently addressing the difficult problem of mapping the location of the solidification boundaries using molten core aluminum analogs of steel alloy systems.

b. Hot EMAT Receiver Design and Testing (PNL)

A complementary effort at PNL concentrated on sensor development (6) and on technology transfer. The primary goal was the development and delivery of a fieldable prototype hot EMAT sensing system. This effort lead to the development of a simple, inexpensive EMAT receiver capable of continuous application on steel samples with surface temperatures in excess of 1300 Celsius.

The EMAT approach is in principle non-contacting but just barely: signal strength decreases exponentially with liftoff distance (Fig. 2). Although liftoff distance of up to 1 cm are possible, a few mm is more usable and still sufficient to reach into the base material through scale and surface irregularities.

To apply ultrasonics using EMATs, the workpiece must be amenable to inspection with conventional ultrasonic equipment (1 MHz range) with no large voids or inclusions. Variations in surface conditions (undulations, heavy scale, etc.) may limit applicability due to reduced signal-to-noise and uncertainty in ultrasonic path lengths.

Recent work at PNL has sought to improve the thermal tolerance of the EMAT designs. This has been accomplished through clever temperature tolerant coil designs, integrated water cooled heat exchangers, ceramic face-plate designs, and modifications to the associated electronics (7). Substantial improvements in heat tolerance, long term survivability, and signal processing are yet possible.

c. **EMAT (EMpulser) Supplants Laser Transmitter**

The original approach at both NBS and PNL used a laser to generate the ultrasonic pulses. This type of pulse is convenient in a laboratory environment, being temporally and spacially precise and highly reproducible. The use of a laser in a mill environment is entirely feasible but adds considerably to the system cost and complexity and obliges special training and safety precautions.

During early testing, a Magnasonics-supplied EMAT receiver was modified to act as a transmitter (8). Generating sharp electromagnetic impulses, this EMAT transmitter (Empulser) successfully launched ultrasonic pulses into the hot steel. Although signal strengths were greatly reduced, this approach was very attractive because it supplanted the laser and considerably reduced the system cost and technical complexity. The reduced signal strengths were combated with refined EMAT and preamplifier designs, and clever signal processing approaches. The EMpulser idea, borrowing much of the technology already developed for the EMAT receiver, was an intriguing option, and recognized as such by our task unit. The combined EMpulser-EMAT receiver was specified for the subsequent field trials.

d. **System Assembly (Hot EMAT - EMAT)**

The prototype EMAT-EMAT system was assembled in shippable cases and tested prior to shipment at PNL. The system included several spare EMAT components, coils, refractory elements, etc. (Fig. 3). Real-time data acquisition and signal processing capability were provided by a Nicolet 4094A digital processing oscilloscope and an H-P plotter. Data could be stored on floppy discs for later analyses with the Nicolet or other computers using appropriate software.

Excellent signal-to-noise ratios were obtained in tests on aluminum reference blocks, and on fine grained (rolled or forged) ferritic steel samples. Poorer signal-to-noise ratios were obtained on coarse grained as-cast samples, obliging special signal processing. This circumstance forewarned that the projected field trials on a horizontal caster would present severe testing conditions.

The control electronics (Fig. 4) included hardware to generate multiple pulses with intermediate data acquisition and storage. This multiple data acquisition approach was needed so that time averaging and other signal processing techniques could be applied to improve marginal signal-to-noise ratios. Requisite signal/noise were obtained in the laboratory on various samples of material at temperatures in excess of 1300 Celsius.

Laboratory testing also demonstrated that pulsed EMATs required judicious practices: on ferromagnetic samples Barkhausen noise caused by magnetic domain motion

had to be eliminated; stable timing circuits were needed to synchronize the three
pulse circuits used in the EMpulser-EMAT system, to adjust timing for different
sample lengths, and to provide an accurate time-zero reference for TOF
measurements.

e. Integration and Field Testing (ARMCO Research & Technology)

The equipment was packaged and shipped to Armco Research & Technology for further
refinement and integration with an actuation system destined for used on the
Baltimore Specialty Steel Corporation (BSSC) horizontal caster (Fig. 5). In-
numerable improvements and revisions were made to the system and performance improved to the point at which useful signals could be obtained on cast stainless
steel samples without multiple shot averaging. Signal-to-noise on samples of the
BSSC material exceed our minimum 3:1 requirement; the ratio on low carbon cast
billets approached 50:1.

The specific aims of this phase of the effort were:

. to engineer the EMAT-EMAT handling and deployment subsystems,

. to correograph operating procedures which avoided damaging equipment and
   experimenters,

. to conduct further hot tests on various materials of interest,

. to fully test all components and spares.

The actuation systems consisted of independently operated fixtures for the EMpulser and EMAT receiver. The EMpulser was located below the sample and articulated
into position by means of an air cylinder; the EMAT receiver was suspended above
the workpiece on the end of a counter-balanced beam. Reflective heat shielding
was used to protect wiring, water hoses, and electronic equipment from heat
radiation.

The EMpulser air cylinder would be activated, gradually bringing the EMpulser into
contact with the bottom of the workpiece; simultaneously, the lever arm would be
manipulated so that the EMAT receiver would contact the top, and the data acquisition and storing sequence would begin. At the end of the acquisition cycle both
EMpulser and EMAT would be retracted. The entire sequence would be repeated at
intervals dependent on the sample temperature.

REAL WORLD TESTING: HORIZONTAL CONTINUOUS CASTER

The equipment was transported to the BSSC test site and staged at a series of
monitoring points progressively closer to the casting machine. The test team included participants from AISI, PNL, Magnasonics as well as support personnel from
BSSC.

In order to complete our commitment to select and oversee the development of candidate technologies it was important to demonstrate that the EMAT sensors could
survive and be effective in a real world environment. Although path length sensors of requisite accuracy are commercially available, none was included in the
plant trials: the emphasis was to demonstrate EMAT survivability and the ability
to detect signals with adequate signal-to-noise. We captured ultrasonic waveforms
at temperature (Fig. 6) and performed elementary signal processing to extract the

TOF data. The billets were interrogated as received and no special surface preparation of any kind was attempted. Some billet dimensions were measured subsequently using mechanical micrometers.

Our initial installation on the horizontal caster was at a location in-line just after the cutoff torch and before the cooling bed. Surface temperatures ranged from 650 to 760 Celsius. The EMAT sensors were placed in contact with the head and tail ends of successive bars momentarily stopped at our inspection station.

Good results were obtained: clear signals were resolved in the as-cast 10 cm X 10 cm product, and the TOF data clearly indicated the expected temperature difference between the front and tail ends of the bars (Fig. 7). Processing and analyses of these data are in progress, and no attempt to reconstruct temperatures has been attempted pending the availability of proper calibration curves. We subsequently tested 15 cm X 20 cm product and found marginal signal-to-noise ratios; off-line signal processing on these stored data was used to extract TOF data.

The system had proven successful at this location, but our goal was to test the limits of that success even to failure of the EMAT device. Accordingly, we moved the installation upstream to within 14 m of the mold. Radiant heat loading on equipment was noticeably higher; strand surface temperatures were in the range of 870 Celsius. Although the EMATs survived intact, electronic equipment, including the Nicolet computer, which were designed to operate primarily under laboratory conditions, began to suffer heat related failures.

Considerable effort was expended in improving the heat tolerances of various components: more heat shielding, relocating the preamplifier, forced air cooling for pulsers and electronic components, a borrowed air conditioner for the main electronics cabinet. These enhancements proved effective and no further heat related failures were experienced.

After adequate TOF data were acquired at the 14 m location, the equipment was reinstalled within 7 m of the mold face: strand surface temperatures were 1100 - 1150 Celsius. Our first test proved unsuccessful: the signal-to-noise ratio was too low to extract usable data. We subsequently determined that this particular alloy (an ARMCO proprietary grade) could not be inspected by conventional ultrasonic methods even at room temperature due to a peculiarly high acoustic attenuation.

We successfully acquired TOF data on the next available heat just prior to plant shutdown; surface temperatures exceeded 1130 Celsius. At the end of cast the bar was repeatedly interrogated as it cooled and the dependence of TOF on bar temperature was clearly resolved (Fig. 8).

SUMMARY

The hot EMAT approach developed to solve this problem is a generic approach to the broad class of sensing problems accessible by ultrasonic techniques. The EMAT approach offers particular benefit for monitoring properties in difficult and marginally accessible locations. Versions of this EMAT might be applied to direct in situ nondestructive evaluation (flaw detection), characterization of material properties, residual stress or texture. Here, the method provides ultrasonic data for measuring the internal temperatures in strand cast blooms without surface preparation.

The hot EMAT-EMAT sensing system proved viable in the field trials on a horizontal continuous caster. The test conditions were a severe test of the EMAT-EMAT approach and we were satisfied that the system survived and usable signals were extracted. Based on laboratory inferences, we anticipate much improved signal-to-noise ratios on as-cast carbon steels and on rolled and forged (fine grained) products.

FUTURE WORK: SYSTEM ENGINEERING CAVEATS

Success of the EMAT-EMAT approach for temperature measurements in other hot charge applications will depend on how effectively the various system components can be engineered for the severe environmental conditions and precise measurement requirements. Key items to be considered include:

. inducing the initial impulse into the sample (EMPulser),

. detecting the pulse upon arrival (EMAT receiver),

. a reliable and accurately aligned transducer handling mechanism,

. path length determination,

. calibration (velocity dependence on temperature),

. rapid real-time signal acquisition,

. signal processing, storage and retrieval,

. algorithmic reconstruction of temperature distribution (for many hot charge applications average temperature may suffice),

. presentation of pertinent results.

We believe that working in concert with NBS and PNL we have made significant progress on each of the requisite components of such measurement systems. Additional effort by PNL and NBS are clearly warranted; our task is to encourage this further development and to promote commercialization of this technology.

ACKNOWLEDGEMENT

The authors gratefully acknowledge the contributions of Don Boyd, Paul Sperline, Gary Spanner, and Doug Lemon (PNL), Haydn Wadley, Steve Norton, and Floyd Mauer (NBS), Don MacLauchlin and George Alers (Magnasonics), the members of the AISI collaborative task unit, and numerous other individuals who supported our efforts.

REFERENCES

The eight AISI member companies participating in the Collaborative Task Unit (5-4) were: Weirton Steel, National Steel Corp., USSteel Division of USX, The Timken Company, Inland Steel Corp, Republic Steel Corp (now part of LTV), Bethlehem Steel Corp., ARMCO Inc.

J. R. Cook, D. F. Ellerbrock, T. R. Dishun, H. G. Wadley, and D. M. Boyd, "Hot Charging and Direct Rolling of Continuous Cast Steel: In-Process Characterization and Control", 1987 Iron and Steelmaking Conference, Steelmaking Conference Proceedings, Pittsburgh, Vol. 70, p. 285f.

R. Mehrabian, R. L. Whitely, B. C. Reuch and H. N. Wadley, Editors, "Process Control Sensors for the Steel Industry" Report of Workshop held July 1982, NBSIR 82-2618, U.S. Dept Commerce.

H. N. G. Wadley, S. J. Norton, F. Mauer and B. Droney, "Ultrasonic Measurement of internal Temperature Distribution," Phil. Trans. Royal Society London, A 320, 341- 361 (1986).

B. E. Droney, F. A. Mauer, S. J. Norton and H. N. G. Wadley, "Ultrasonic Sensors to Measure Internal Temperature Distributions" Review of Progress in Quantitative Nondestructive Evaluation, edited by D.O. Thompson and D. E. Chimenti, (Plenum Press, New York, 1988 Vol. 5A, pp. 643-650.

Subcontractor Magnasonics, Inc. (Albequerque, NM) developed the original pulsed EMATS and Empulsers used during the laboratory testing.

D. M. Boyd and P. D. Sperline, "Non Contact Temperature measurements of Hot Steel Bodies Using an ElectroMagnetic Acoustic Transducer (EMAT)", Review of Progress in Quantitative Nondestructive Evaluation, edited by D. O. Thompson and D. E. Chimenti, (Plenum Press, New York, 1988), Vol 7B, 1669-1676.

L. R. Burns, G. A. Alers, and D. T. MacLauchlin, "A Compact EMAT Receiver for Ultrasonic testing at Elevated Temperatures", ibid, 1677-1683.

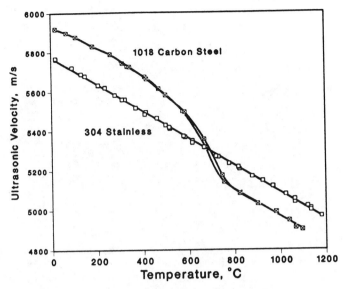

Fig. 1 Longitudinal ultrasonic velocity dependence on temperature. A nearly linear dependence is found with austenitic 304 stainless steel. A more complex dependence with noticeable hysteresis is found on ferritic 1018 steel.

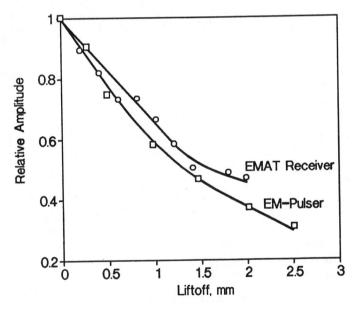

Fig. 2 EMAT receiver and EMpulser sensitivities show exponential dependences on liftoff distance. Liftoffs of up to 1 cm can be used, but a few mm is more practical and usually sufficient to overcome surface irregularities and scaling conditions.

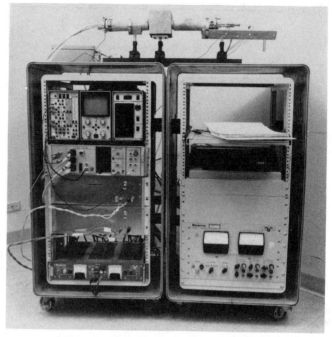

Fig. 3   Photograph of the PNL EMAT receiver and EMAT pulser just prior to shipping to Armco Research. Also shown are the power supplies, control and data acquisition equipment mounted in shippable containers for field testing.

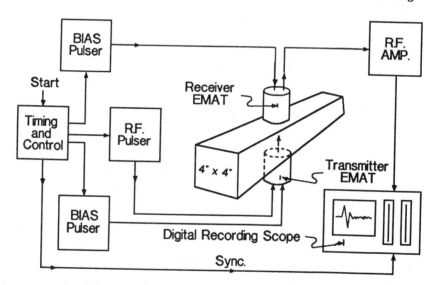

Fig. 4   Block diagrams of signal and timing blocks for EMAT-EMAT signal detection and storage. Three separate pulsing circuits with associate time controllers are needed.

Fig. 5 Photographs of the Armco Research laboratory arrangement showing the EMpulser and EMAT receiver delivery systems in the data aquisition position on a reheated billet.

Fig. 6 Photograph of the EMAT Receiver and EMAT pulser contacting a 10cm × 10cm stainless steel billet in Baltimore Specialty Steel Corporation horizontal continuous caster. Some billets were marked for later verification of dimensions.

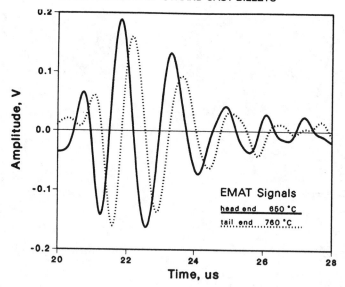

Fig. 7   Data acquired in sequential tests of the leading and trailing ends of one
10cm×10cm cast billet. The shift between traces indicates a difference of more
than 100 Celsius in internal temperature between front and tail of the bar.
Measured surface temperatures were 650 and 760 Celsius.

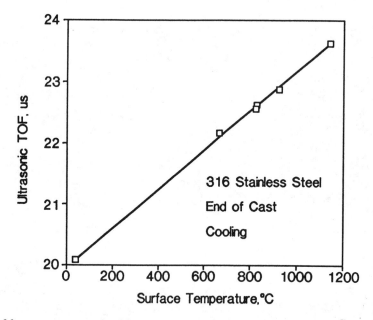

Fig. 8   Cooldown curve acquired by repetitively interrogating last bar produced at the
end of cast. Ultrasonic transit times sensitive to the internal temperatures are
plotted against surface temperatures measured by handheld pyrometer. The
dependence is surprisingly linear.

Mathematical Models and Sensors as an Aid to Steel Quality Assurance for Direct Rolling Operations

R.I.L. Guthrie, S. Joo, *H. Nakajima

McGill University, Montreal

*Sumitomo Metals, Kashima Works, Japan

INTRODUCTION

Direct rolling of slabs represents a further hurdle in the quest for continuous steelmaking. Energy savings associated with direct rolling are significant and in the order of 0.5 GJ/tonne. These represent a 3.6% decrease from current best steel work practices of 13.8 GJ/tonne (from pellets to hot band) and bring steelmakers one step closer towards the minimum thermodynamic value of about 6.5 GJ/tonne for the production of steel from iron ore. These savings are brought about by the elimination of inspection procedures on cold slabs and subsequent slab reheating.

For these benefits to be realised, the quality of steel entering hot mills must be assured in terms of chemistry, temperature and other physical characteristics, such as integrity (freedom from cracks, center-line porosity, scale, etc.) and metal cleanness (minimum levels of exogeneous inclusions).

The present paper is concerned with describing (a) recent developments at McGill University in the application of computational fluid dynamics (CFD) to the optimisation of inclusion removal and temperature control in the tundish of the Stelco Hilton Works Single Slab/Twin Bloom Caster arrangement and (b) concurrent development of an inclusion sensor for steel melts. The twinning of these two subjects rests on the logic that mathematical models have now been developed to the point that they can be used in the design of reliable tundish configurations for optimising thermal, chemical and inclusion control, but that sensors for temperature, chemistry and inclusions are still needed to monitor inevitable departures from any optimised, set-point levels. Such departures are either adventitious (e.g. ladle slag vortexing) or scheduled, such as ladle and tundish changes during sequence casting.

Fluid Modelling of Flows in Tundishes

Through solution of the turbulent Navier-Stokes and continuity equations, plus appropriate prescriptions for turbulence, steel flows generated within the tundish and mould of a slab casting machine are now predictable in a quantifiable sense. Similarly, associated thermal phenomena, such as the various temperature drops experienced by steel flowing from an inlet nozzle to the various outlet nozzles can be deduced, given appropriate information on wall heat losses, etc. The differential equations describing such phenomena have been documented in previous publications (e.g. Ref. 1) and are therefore not repeated. Obviously the boundary conditions still remain, and are model, or steel-plant, specific. A general purpose finite-difference code, based on the SIMPLE algorithm, was developed for the description of single phase, steady or transient, three dimensional flows in rectangular and/or cylindrical vessels. Most of the computer runs described were carried out on a desk-top microprocessor (IBM-AT) fitted with a hard disc and a Definicon System of 8 MB RAM and 20 MHz clock speed. Typical 3D runs using grids of 17 x 36 x 16 required about 6 hours CPU time for flow field computations.

TABLE 1   IMPORTANT PARAMETERS AND PROPERTIES

OF MODEL AND PROTOTYPE

		Model	Prototype
Geometry	Tundish Length	5.19m	5.19m
	Tundish Depth	1.10m	1.10m
	Bottom Width	0.68m	0.68m
	Surface Width	1.07m	1.07m
Fluid Properties	Liquid	Water	Steel
	Temperature	15°C	1550°C
	Density	1000 Kg/m3	7000 kg/m3
	Viscosity	1.14×10^{-3} kg/m3	6.7×10^{-3} kg/m.s
	Volumetric Flowrate	6.9×10^{-3} m3/s	6.9×10^{-3} m3/s
Inclusion Properties	Inclusions	Glass microspheres	Al_2O_3 and/or SiO_2
	Size Range	20~110μm	–
	Density	295 kg/m3	≃3000 kg/m3

Thus Fig. 1 predicts flow patterns that will be generated in a half section of the new Hilton Works tundish, when a dam and weir arrangement is placed one-third way along the length of the tundish slab casting. Key dimensions and flowrates are provided in Table 1. Figure 1A presents an isometric view of the flows generated, and these may be interpreted in terms of Fig. 1 B, where a collage of two dimensional plots of velocity components along selected longitudinal and tranverse axes of the tundish are given. This particular tundish has sloping sidewalls of 10°. Sloping sidewalls may act (in part) as an aid for inclusion float up, accumulation and possible separation into an absorbent slag. However, increased radiative heat losses can be expected from such designs, in the absence of tundish covers or slag blackets.

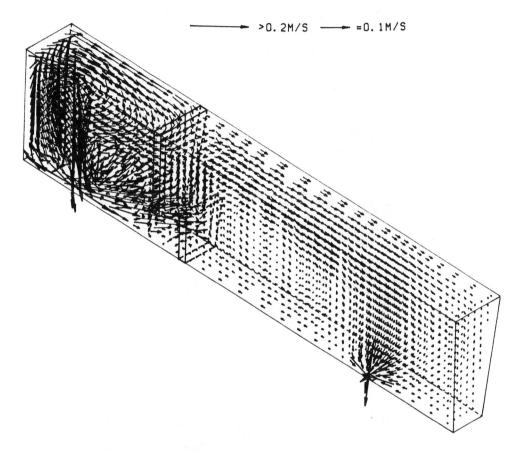

Fig. 1A   An isometric view of flow fields predicted in the longitudinally bisected single strand tundish of slab casting Stelco's Hilton Works.

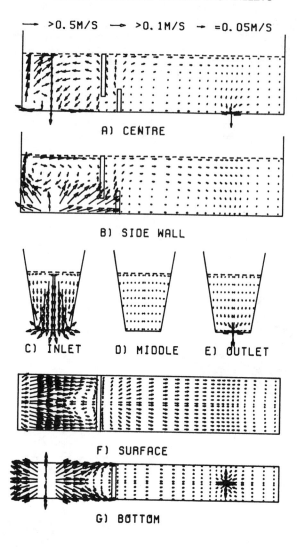

Fig. 1B   Predicted flow fields in some selected longitudinal planes (a,b), transverse planes (c,d,e), and horizontal planes (f,g) for the single strand slab casting arrangement.

Without dwelling upon the obvious complexities in the flows induced by the presence of dams and weirs, it is clear that previous and often current practices of modelling tundishes in terms of well mixed, plug, and dead flow regions represents a useful, but gross, simplification of real events. While the judicious choice of relative proportions of mixed and plug flow regions can mimic typical experimental mixing curves, they are in no sense predictive, and need to be restricted to simple systems.

Take for example Fig. 2, showing equivalent flows when the tundish is being operated in the double bloom mode; clearly, different 'well mixed, plug and dead flow zones' are needed to predict inclusion separation behaviour at the two nozzles exits. Such postulated zones would be typically deduced from tracer response curves at the two outlets.

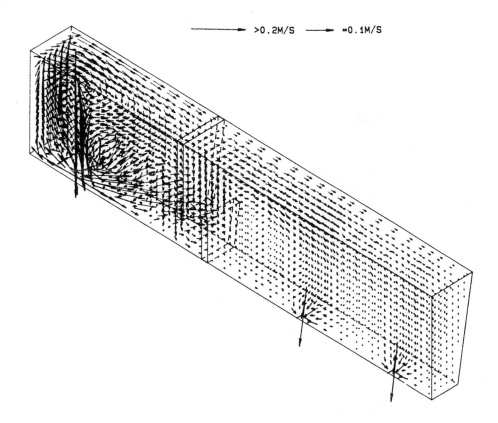

Fig. 2A   An isometric view of flow fields predicted for a
half volume of the tundish set for twin bloom
casting.

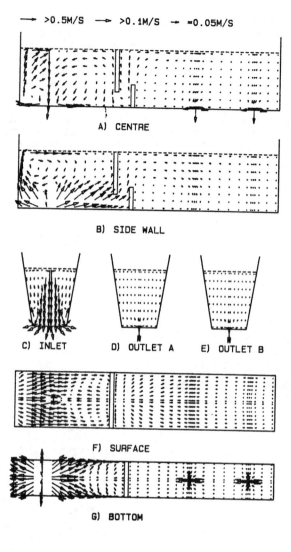

Fig. 2B   Predicted flow fields in selected longitudinal
           planes (a,b), transverse planes (c,d,e) and
           horizontal planes (f,g) of the tundish set for
           twin bloom casting.

Figure 3 demonstrates the relative insensitivity of mixing times on dam and weir placements in a large deep tundish. Shown are the tracer response curves which suggest that a dam and weir placed one third way down the tundish should give about the longest "minimum residence time" in the slab cast mode. However, these curves suggest no radical improvement in minimum residence times over an unobstructed tundish. (Dramatic differences in these curves can however be achieved for tundish with highly sloping sidewalls).

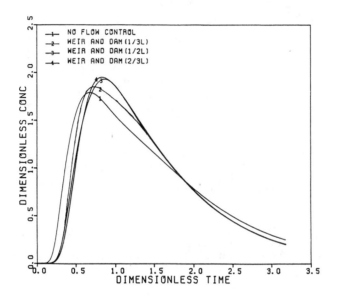

Fig. 3   Residence time distribution (RTD) curves predicted with, and without flow controls, for slab casting arrangements.

Corresponding particle/inclusion separation curves for these dam and weir arrangements are shown in Fig. 4. As seen, the Residual Ratio (i.e. those inclusions still present in the effluent stream expressed as a fraction of those entering) of very small inclusions is unity at all dam/weir combinations, and zero, for all large inclusions. This is to be expected, since very small particles will have minimal Stokes rising velocities and are therefore unable to separate, while inclusions with rising velocities in the order of 5-6 mm/s will have an adequate opportunity to accumulate in the top regions of the tundish, given a mean tundish residence time of 7.8 minutes. The benefits of a dam/weir system for the Hilton system are to be had for particles of intermediate Stokes rising velocities (1-5 mm/s). These values are naturally specific to inclusion type, tundish design and flowrates. Computations show that a dam/weir arrangement placed at 1/3L will give the best inclusion separation ratio (i.e. minimise the Residual Ratio).

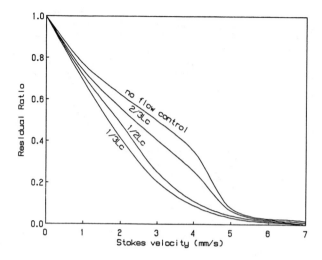

Fig. 4   Relationship between the residual ratios of inclusion particles and Stokes rising velocities predicted for a full scale water model of the slab casting tundish.

For a multi-ported tundish, such a choice would be less obvious since the quality of steel exiting the inside port will tend to be poorer than that exiting the far port, there being less time for float-out. Fig. 5, for the twin bloom casting arrangement illustrates this point.

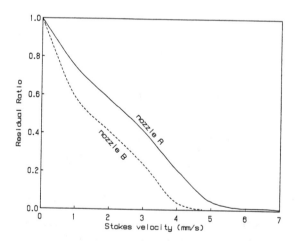

Fig. 5   Relationship between the residual ratios of inclusion particles and exit nozzle of tundish set for twin bloom casting.

Comparison of inclusion residual ratios with experiment is provided in Fig. 6. There, the residual ratios of hollow glass micro-spheres (simulating inclusions rising through steel) were monitored 'on-line' using resistive pulse detection equipment originally developed at McGill by S. Tanaka and described in detail elsewhere [2]. As seen, good agreement was achieved between predicted and measured R.R. values. It is emphasized that this data refers to *quasi steady state* conditions, following the continuous injection of inclusions into the tundish. This is typically achieved after about three mean residence times, and corresponds to those conditions wherein inclusions within the recirculating flow zones have accumulated to steady levels, while those in stagnant zones continue to accumulate. These computations refer to isothermal systems.

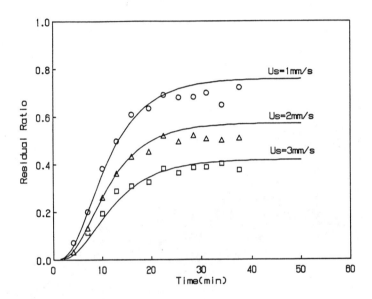

Fig. 6   Comparison of experimentally measured and predicted particle separation ratio versus time of casting for the single strand tundish with no flow control.

In real tundishes, one can observe significant drops in metal temperature between entry and exit points (4). Consequently, the possibility of thermal natural convection currents causing a modification to flow patterns is a distinct possibility that mathematical and physical modellers have neglected until now.

Figures 7A and 7B present a typical set of computations wherein <u>forced</u> <u>and</u> <u>natural convection</u> of steel within the tundish were taken into account. In modelling the heat losses from the steel to the surrounding brickwork/liner board sidewalls and base, heat fluxes corresponding to radiative and natural convective heat losses from 200°C exterior sidewalls were applied. Similarly, in estimating surface heat losses from the steel, it was assumed that these were radiative in nature, with no slag or tundish covers present. Under such conditions, one can see that the flow patterns are markedly changed from those in Figure 1A, for isothermal conditions. The main difference is a much stronger flow down the side and particularly end wall of the tundish, coupled with a stong flow along the bottom surface and much stronger zone of recirculation to the right of the dam and weir arrangement. While this is beneficial for temperature uniformity, the greater departure from plug flow conditions will have a detrimental bearing on steel quality (viz inclusions). Other changes resulting from natural convection include a stronger recirculating zone in the entry region, with reversal in the surface flows back towards, rather than away, from its ladle shroud entry port.

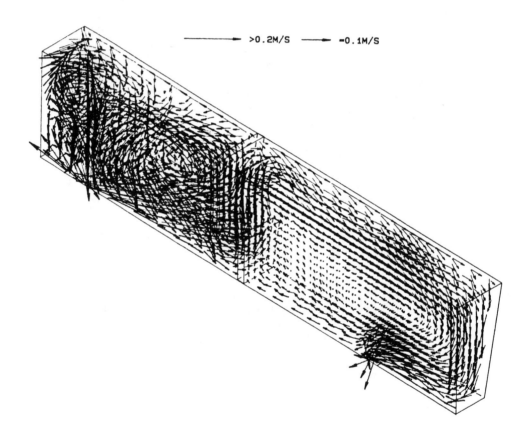

Figure 7A An isometric view of flow fields predicted in the longitudinally bisected tundish for Stelco's Hilton Works single slab/twin bloom caster, taking into account natural and convective heat transport processes (compare Figure 1A)

Figure 7B presents the corresponding isotherms within the steel at longitudinal vertical planes at the central and mid planes, and adjacent to the tundish sidewall. As seen, the net drop in temperature for the conditions modelled was about 40°C, the jet entering at 1600°C and exiting 1560°C. This figure is somewhat higher than typical practices which may vary between 10-20°C depending on throughput, slag cover and tundish lining.

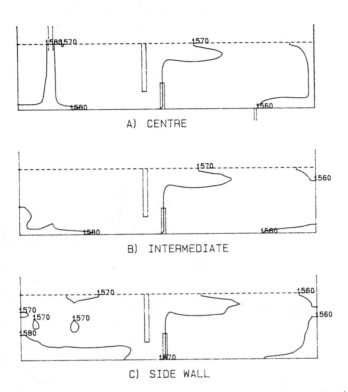

A) CENTRE

B) INTERMEDIATE

C) SIDE WALL

Figure 7B   Predicted temperature isotherms in the Hilton Works caster assuming radiation from an exposed upper surface of steel, and steady state side-wall heat losses to the environment.

One advantage of 'allowing' natural convection to take place is the elimination of the dead zones at the tundish end walls where the propensity for freezing is greatest. Referring to Figure 7B, the 1560°C isotherm at the end wall shows that freezing along the top surface and bottom close to the nozzle is most likely for the particular design of tundish under investigation.

In conclusion, since such heat transfer phenomena are not readily modelled with water, mathematical models present the only reasonable alternative to adequate tundish design, if the importance of ladle covers, slag covers, bubbling,etc., on the design and operation of such vessels are to be properly predicted.

Inclusion Sensors

While mathematical models are essential to the design of optimised metal delivery systems for slab casters, sensors are needed to record excursions from pre-set levels. Following the successful development of a probe for monitoring inclusions in liquid aluminum (5), work has been pursued at McGill, aiming to repeat this success in the far more hostile environment of liquid steel at 1600°C. One of the authors, Dr. H. Nakajima of Sumitomo Metals Industries, spearheaded work on the development of a practicable continuous probe suitable for steel melts. Following laboratory tests, preliminary plant trials were carried out at the McMaster Works of Stelco, and some of the results of these tests for silicon killed low carbon steel grades are now presented. Thus Fig. 8A shows the probe inserted in the T shaped tundish of Stelco's four strand casting machine, while Fig. 8B shows the probe following fifteen minutes of immersion.

Fig. 8A   Photograph of the LIMCA quartz probe system used for molten steel tests in the tundish of a four strand casting machine.

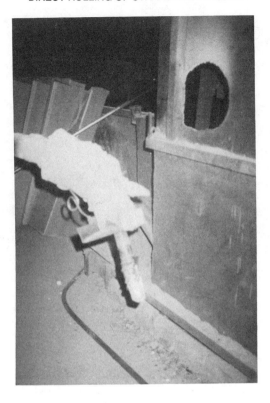

Fig. 8B   Limca probe after being submerged in the
            1550°C melt for 15 minutes.

The method relies on sensing inclusions passing through an electric sensing zone. Thus, when particles disturb an electric field within a metal, the slight change in electrical resistances caused by their presence, can be detected provided high D.C. currents of electricity, and major signal amplification (~1000x), are practised (6). The most convenient way of establishing this electric sensing zone (E.S.Z.) to date, is to draw a sample of the melt into an insulating tube, via an orifice (the E.S.Z.), and to record the voltage pulses corresponding to the passage of *individual* particles. Figure 9A gives a typical Limca pulse monitored during this sampling, indicating the passage of an inclusion through the E.S.Z. Figure 9B shows the sensing zone aperture which is set in the lower portion of the sampling tube, before exposure to the melt and Fig. 9C, after use. As seen some pitting of the orifice is evident, but dimensional integrity was maintained.

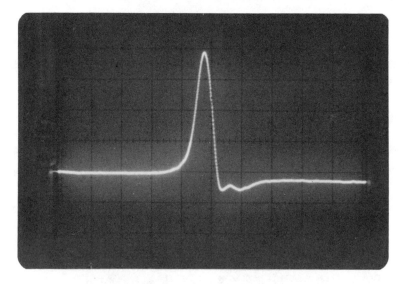

Fig. 9A   Voltage pulse obtained during LIMCA measurements in a T tundish used for continuous billet casting (Stelco, MacMaster Works, 4 strand tundish).

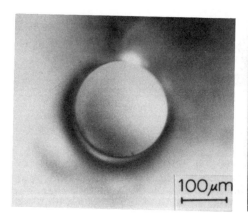

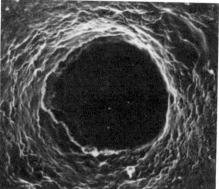

Fig. 9B   Magnified view (100x) of area surrounding the quartz orifice prior to an immersion test.

Fig. 9C   View of quartz orifice following 10 minutes continuous operation in molten steel held at 1550°C.

Figure 10 shows a size distribution analysis deriving from a twenty second sampling of the melt, while Fig. 11 was made possible by taking Limca readings at the entrance and exit regions of the T tundish. Thus by analysing inclusion frequency size distributions at the exit to those at the entrance, in a similar fashion to that for water models (7), steel plant data for residual ratios (R.R.) can be plotted. The Limca data points correspond to the larger particles in Fig. 11, while microscopic analyses of solidified melt samples taken at the entry and exit yielded the open-circled points. The relationship between ln (RR) and U_s (the Stokes Rising Velocity) coincides with the type of curves observed from water models. While the reliability and use of the technique by the steel industry is well behind its use by aluminum producers, it is worth illustrating its potential for quality control by showing its application to **Alcan's** aluminium can body grades. (Fig. 12). There the inclusion content of molten aluminum is continuously monitored by Limca during the casting of D.C. ingots; the section delivered to the customer has to fall within appropriate levels of metal quality.

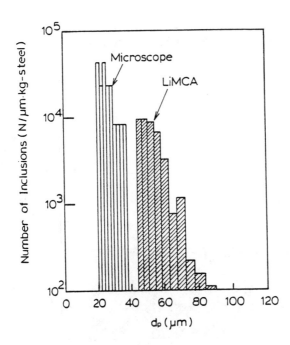

Fig. 10   Inclusion size distribution measured in a tundish using the Limca method, together with that obtained from the microscopic analysis of a 'lollipop' sample.

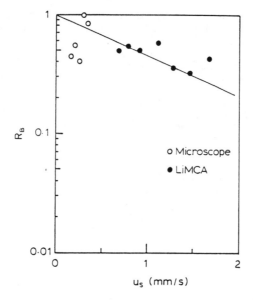

Fig. 11   Plot of logarithm of an inclusion's residual ratio
            versus its Stokes rising velocity (mm/s).

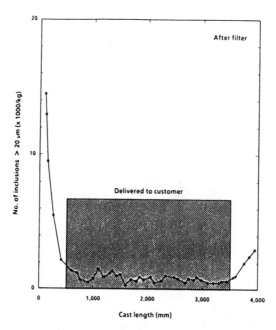

Fig. 12   Plot of metal quality in a D.C. cast aluminum
            ingot versus cast length, showing the section
            delivered to the customer. Metal quality
            defined as Number of inclusions greater than
            twenty microns (x1000) one kilogram of metal.

CONCLUSIONS

The elimination of slab inspection and reheating furnaces requires that metal from the caster be paced to hot mill production schedules and be of prime quality. Mathematical models of liquid steel flows in transfer vessels can provide a powerful tool in the design of metal delivery systems for optimum steel quality (temperature, chemistry, inclusion levels). Sensors, such as the Limca technique, and continuous oxygen probes, could prove to be invaluable accessories to monitor any metal quality excursions during actual slab casting and rolling operations. Such sensors could be placed within the mould to monitor metal quality from the jets issuing from the submerged entry nozzles, or other strategic locations.

ACKNOWLEDGEMENTS

The authors would like to acknowledge the support of the Natural Science and Engineering Research Council in supporting the inclusion sensor work via a C.R.D. grant, and the mathematical modelling work via a strategic grant. Similarly, they acknowledge the enthusiastic help and support of industrial colleagues; in particular Don. J. Harris, Jack D. Young, Lucy Dumitru, Kevin O'Leary and Angelo Grandillo, all of Stelco Inc.

REFERENCES

1. M. Salcudean, C.H. Low, A. Hurda and R.I.L. Guthrie, Chem. Eng. Comm., Vol. 21, 1982, pp. 89-103.
2. S. Tanaka, "Modelling inclusions behavior and slag entrainment in liquid steel processing vessels", Ph.D. thesis, McGill University, 1986.
3. F. Sebo, F. Dallaire, S. Joo and R.I.L. Guthrie, "On-line sizing and monitoring of particles suspended in aqueous solutions", Int'l Sym. on Production and Processing of Fine Particles, Paper 12-3.
4. Y. Maruki, S. Shima and K. Tanaka, Transactions ISIJ, Vol. 28, 1988 pp.
5. D.A. Doutre, R.I.L. Guthrie, "On-line measurements of inclusions in liquid melts", Intl'l Seminar on Refining and Ferro-alloys, Ed. T.A. Engh. S. Lyng, H.A. Oye, Aluminum-Verlag Dusseldorf, pp. 147-163.
6. D.A. Doutre, R.I.L. Guthrie, "An apparatus for the detection and measurement of particles in liquid metals", U.S. Patent 4,555,662, Nov. 26, 1985.
7. H. Nakajima, "On the detection and behavior of second phase particles in steel melts", Ph.D. thesis, 1986, McGill University, Montreal.

SESSION 4

HCR AND DR PROCESSING

Chairpersons: G.A. Irons (McMaster University)
                      S. Yue (McGill University)

APPLICATION OF THE HCR PROCESS IN BLOOM CONTINUOUS CASTING

Ryuzo Komatsu and Kenzo Yamaguchi

Technical Sect.-Melting   Production Engineering Dept. No.1

AICHI STEEL WORKS, LTD.

ABSTRACT

The bloom caster at Aichi Steel Works produces 610 kton/year of Al-killed steel and was installed in 1982 with the view to applying the Hot Charge Rolling (HCR) process. 90% of the blooms produced are rolled through the HCR process with significant energy savings when subsequently charged to the reheating furnace. The HCR process requires the supply of blooms which are free of surface defects and of sufficiently high temperature for direct charging. The technical aspects of achieving this are;
1. Prevention of macroscopic surface inclusions by supplying clean molten steel treated in LF-RH process, shrouding of molten steel, floating up alumina inclusions in tundish.
2. Prevention of surface cracks by decreasing the secondary cooling water and adoption of the Bloom Dipping Bath Process.
3. Transportation of blooms by heat insulating bogie cars.

INTRODUCTION

No.1 bloom caster, which is installed beside the established blooming mill, started production in November 1982. At present, No.1 bloom caster produces 610 kton/year of Al-killed steel with 90% of blooms produced rolled through the HCR process. This paper reports the features of the No.1 bloom caster, the production techniques of supplying blooms free of surface defects and of sufficient temperature, and the HCR operation results of No.1 bloom caster.

THE CHARACTERISTICS OF NO.1 BLOOM CASTER

The main specifications of No.1 bloom caster are shown in Table 1 and the layout of electric furnace, No.1 bloom caster and blooming mill is shown in Fig. 1. The characteristics of No.1 bloom caster for applying the HCR process are as follows;

1. Elimination of non-metallic inclusions.
   a) Treatment of all molten steel in LF-RH process.
   b) Shrouding molten steel from ladle to mould.
   c) Adoption of a large capacity tundish.
2. Improvement of bloom surface quality.
   a) Adoption of automatic mould level control.
   b) Decreasing secondary cooling water.
   c) Bloom Dipping Bath process.
3. Transportation of blooms with high temperature to the reheating furnace using heat insulating bogie cars.
4. Automatic operation of the continuous casting process and the computer transfer of bloom data information to the blooming mill shop.

PRODUCTION TECHNIQUES FOR THE ELIMINATION OF BLOOM SURFACE DEFECTS

There are two types of defects on the surface of Al-killed blooms. One is macroscopic surface inclusions which occur due to trapping alumina inclusions and mould powder into the solidified shell. The other is surface cracks which occur due to non-uniform secondary cooling and excessive deterioration of surface temperature.

Prevention of Macroscopic Surface Inclusions

Treatment of molten steel in LF-RH process.[1,2] In order to reduce alumina inclusions in molten steel, clean molten steel which is treated in LF-RH process (as shown in Fig. 2.) is supplied to No.1 bloom caster.

Shrouding molten steel from ladle to mould and a large capacity tundish. It is very important to prevent re-oxidation of clean molten steel and to float up inclusions in the tundish effectively. For this purpose, a shrouding system involving long nozzle and submerged nozzle is used in conjunction with a large capacity tundish. The tundish of 20T weight and 800mm depth is fitted with triple dams. The surface inclusions on the initially cast bloom can be further reduced by purging the tundish and the mould with argon gas, overlapping the triple dams and prolonging the preheating time of the tundish.(Fig. 3.)

Automatic mould level control. If the steel level in the mould fluctuates greatly, forming of the solidified shell and mould powder inflow are changed so that mould powder is trapped into the solidified shell. In order to control mould level effectively, an automatic mould level control system was adopted which consists of a slide valve and mould level detection system.(Fig. 4.) In order to maintain initial molten steel depth in the tundish and start casting smoothly, the steel level in the mould is controlled by manual stopper at the start of casting and then replaced by the automatic mould level control system. To prevent mould level fluctuation due to nozzle clogging, argon gas is blown into the submerged nozzle of which the diameter is 70 mm. By this control system, the steel level fluctuation in mould is controlled o less than 10mm as shown in Fig. 5.

Using these techniques, macroscopic surface inclusions have not occurred in the No.1 bloom caster.

Prevention of Surface Cracks

Decreasing secondary cooling water. Generally speaking, a longitudinal crack and a traverse crack occurs as follows. If formation of the solidified shell is irregular, thermal stress is generated. Due to this thermal stress, a tiny cracks on the solidified shell in the mould occurs and it is expanded by secondary cooling to become a longitudinal crack. If the strand is straightened by pinch rolls through the poor ductility range of 700 °C and 900 °C in which ferrite is precipitated through austenite-ferrite transformation, a traverse crack occurs.

When No.1 bloom caster started up, the secondary cooling intensity was high (specific water ratio is 0.6 l/kg-steel, secondary cooling zone length is 12.6 m) and the strand was straightened at about 800 °C. As the results, tiny longitudinal cracks and traverse cracks occurred on the surface of the blooms.

In order to prevent these cracks, secondary cooling water was decreased by gradually shortening the secondary cooling zone length. Within a specific water ratio of 0.2 l/kg-steel, surface cracks of rolled billets are greatly reduced. (Fig. 6.) However, when the secondary cooling intensity is lower than 0.2 l/kg-steel, the surface cracks persist. Such a situation does not prevent surface cracking of low carbon Al-killed steel to the same extent as middle and high carbon Al-killed steel. The surface temperature change of the strand is shown in Fig. 7.

Bloom dipping bath process. AlN is precipitated in ferrite at the austenite grain boundaries in the ferrite-austenite phase state. If such blooms are heated and rolled, surface cracks occur on the rolled billets because the workability is inferior to blooms in which AlN is not precipitated.

In the HCR process, the bloom surface temperature of low carbon Al-killed steel stagnates in this double phase state giving rise to the surface cracks on the rolled billets. There are two methods to prevent AlN precipitation at austenite boundaries. One is to put blooms more than 850 °C into the reheating furnace before precipitation of AlN. The other is to put blooms into the reheating furnace after cooling them to less than 700°C. By this cooling, coarse grains are refined through transformation and AlN is not precipitated at the austenite grain boundaries.(Fig. 8) It is very desirable to put all blooms greater than 850°C into the reheating furnace, but, it is difficult to keep blooms at such a high temperature under varied casting conditions. Conversely, it is a large energy loss to put blooms less than 700 °C into the reheating furnace.

Other methods, therefore, were investigated in which hot blooms can be charged to the reheating furnace after cooling the surface of the blooms below Ar_3 transformation. Two types of cooling methods were tested. The first was to cool by high intensity spray before solidification. The second was to dip the hot blooms into a water bath after cutting. In the case of spray cooling, internal cracks occurred because of reheating after cooling. Consequently, the dip cooling method after cutting was adopted.(Bloom Dipping Bath Process) The treatment method of Bloom Dipping Bath Process is shown in Table 2 and the bloom surface temperature change is shown in Fig. 9.

This cooling method requires a flow rate of 3000 l/min water into the bath. The drain, not unexpectedly, contained a high level of scale deposits. Consequently, a water circulation system was adopted in which water, after cooling the strand

and machine, is pumped up from the scale pit and is put into the bath and with the drain returning into a scale pit. The blooms in the Bloom Dipping Bath process are handled by automatic crane which is computer controlled. The blooms are pulled into an insulating bogie car after the Bloom Dipping Bath process. Treatment in Bloom Dipping Bath process decreases surface cracks of low Al-killed steel so that it is possible to omit the conditioning of rolled billets.(Fig. 10.)

HIGH TEMPERATURE BLOOM PRODUCTION TECHNIQUE

For bloom transportation from No.1 bloom caster to the reheating furnace, we adopted bogie cars system similar to that used for ingots. It takes about one hour after cutting in No.1 bloom caster to charging into the reheating furnace. In order to minimize the bloom temperature drop during transportation, 10 hot blooms are piled in two lines in an insulating box which is installed on bogie car.(Fig. 11.) The main specification of the bogie car with the insulating box is shown in Table 3. By using these insulating bogie cars, a bloom surface temperature greater than 600 °C can be obtained just before the reheating furnace. As the results, 15000 kcal/ton energy of the reheating furnace can be saved by this HCR process compared to HCR process of ingots.

RESULTS OF HCR OPERATION

By exploiting various techniques of continuous casting, No.1 bloom caster has been operating smoothly since it started up in November 1982. The production of No.1 bloom caster and the hot charged rolling ratio are shown in Fig. 12. and the conditioning ratio of rolled billets is shown in Fig. 13. Production in 1987 is 610kton with a hot charged rolling ratio of 90% and the conditioning ratio of rolled Al-killed billets is 10%. This HCR process makes a great contribution to saving energy of reheating furnace and omitting conditioning of rolled billets.

CONCLUSION

By exploiting the production techniques of supplying blooms without surface defects and with high temperature, No.1 bloom caster has operated HCR of Al-killed steel steadily. This HCR process has a significant effect on saving energy and improving rolled billet conditioning. We intend to exploit production techniques of higher temperature blooms with no defects and to apply Hot Direct Rolling (HDR) process of Al-killed steel.

REFERENCES

1. T.Yamada, J.Eguchi, S.Takaha, T.Futamura, Y,Takada, T.Itoh (1986). Abstract of IISC (Washington), 366
2. T.Itoh, K.Kamo, T.Yamada, S.Takaha, T.Futamura (1986). 2nd European Electric Steel Congress, Vol.1, R4.1.1-R.4.1.14

TABLE 1 Main Specification of No.1 Bloom Caster

Item	Specification
Steel making furnace	80ton EF (Max.143ton/heat)
Secondary refining	VSC, LF, RH
Type of caster	Bow type with two points unbending (Radius : 16.0m/30.4m)
Number of strands	2
Kinds of casting steel	All kinds of low alloy steel
Casting size	Thickness : 370mm Width : 480mm
Ladle exchange	Ladle turret
Tundish capacity	20ton weight, 800mm depth
Mould level control	RI + sliding nozzle, stopper
Mould oscillation	Stroke : 11mm Cycle : Max.200c.p.m.
Secondary cooling	Spray cooling
EMS	S-EMS (two stages)
Cutter	Gas torch cutter
Start up	November 1982

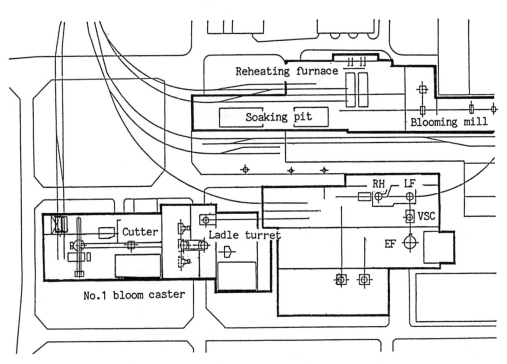

Fig. 1. Layout of electric furnace, bloom caster and blooming mill

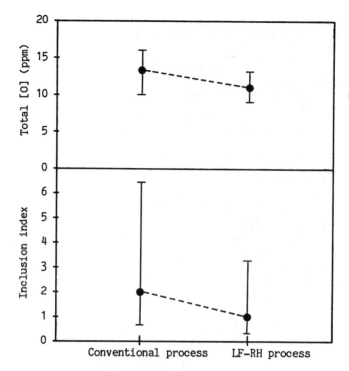

Fig. 2.   Effect of LF-RH process on total [O] and inclusion level

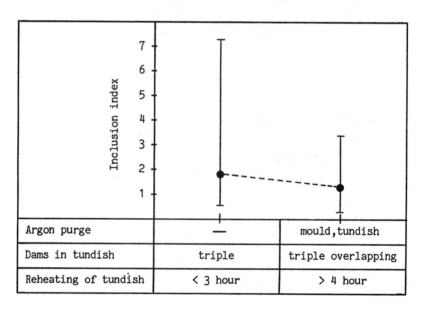

Fig. 3.   Effect of argon purge, overlapping dams and preheating time
of tundish on initial bloom inclusion level

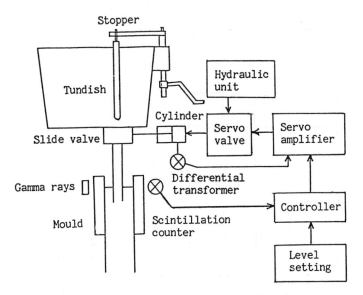

Fig. 4. Schematic diagram of mould level control system

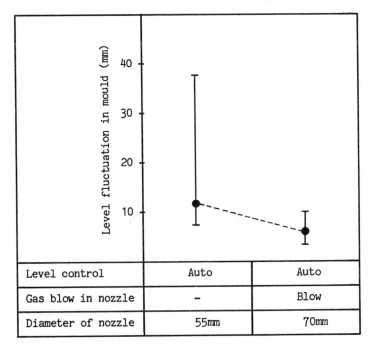

Level control	Auto	Auto
Gas blow in nozzle	–	Blow
Diameter of nozzle	55mm	70mm

Fig. 5. Effect of nozzle clogging prevention
on level fluctuation in mould

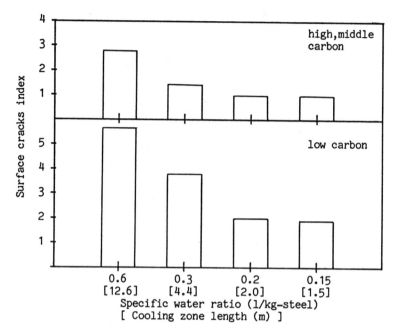

Fig 6. Influence of secondary cooling to surface cracks

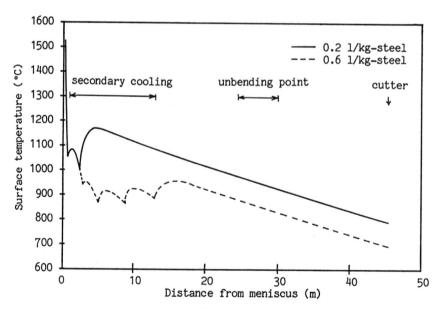

Fig. 7. Change of bloom surface temperature

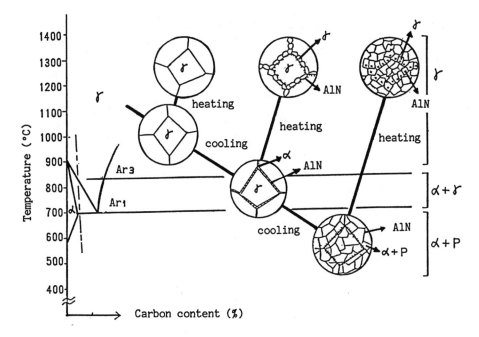

Fig. 8. Schematic view of microstructure due to thermal history

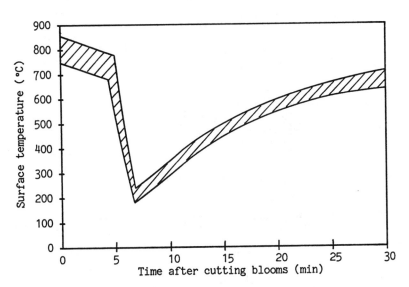

Fig. 9. Change of bloom surface temperature
         through Bloom Dipping Bath Process

TABLE 2 Method of Bloom Dipping Bath Process

Cooling method	Dipping blooms into water
Dipping time	2 minutes
Water temperature	Max. 40 C
Water volume	3000 l/min
Bath size	Length : 4.6m Width : 2.4m Depth : 1.0m
Transfer equipment of blooms	Automatic crane controlled with computer

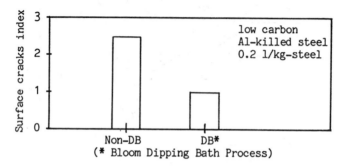

Fig. 10. Effect of Bloom Dipping Bath Process

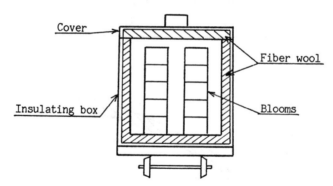

Fig. 11. Schematic view of insulating bogie car

TABLE 3 Main specification of insulating bogie car

Item	Specification
Capacity	70 ton
Structure	Insulating box with cover on bogie car
Insulating box size	Length : 5.2m Width : 2.8m Hight : 2.8m
Insulating material	Fiber wool (thickness : 100mm)

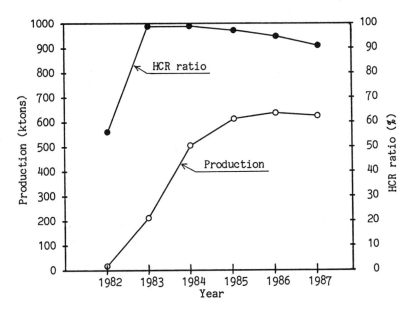

Fig. 12.   Operation results of No.1 bloom caster

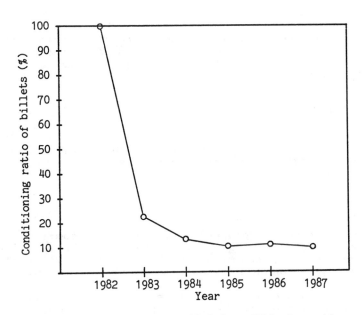

Fig. 13.   Change of billets' conditioning ratio

DIRECT CHARGING OF STRAND CAST BLOOMS
AT ALGOMA STEEL

R.F. Jenkins

ABSTRACT

A review of recent changes made in the steelmaking and bloom casting facilities to allow direct charging of strand cast blooms at Algoma Steel is given. Performance with direct charging and without is compared.

KEYWORDS

Steel; continuous casting; blooms; direct charging; clean steel; submerged entry nozzles; tundish slide gates; water modelling.

INTRODUCTION

Algoma Steel has been a pioneer in the field of continuous casting, having cast blooms since 1969. This paper will review the changes made recently at Algoma Steel, in our No.1 Steelmaking complex that have made direct charging of blooms possible.

For many years, cast blooms were taken from the caster to an inspection area where they were allowed to cool down. After the blooms were cold, they were inspected for defects such as cracks, subsurface porosity and entrapped scum. Defects were laboriously conditioned out. The blooms were reheated and then rolled into product. This was an expensive, energy inefficient and time consuming process.

It was Algoma Steel's goal to produce blooms that were defect free, that could be taken directly from the caster, still hot, and charged into the mill reheat furnace. We refer to this as direct charging. This method would save time, energy and costs and has now been achieved with no sacrifice to quality. In fact our quality has improved with less rejects than ever before, with state of the art inspection equipment, in our newly upgraded rail finishing facilities.

To achieve this, Algoma Steel embarked on a "clean steel" program. At Algoma Steel, "clean steel" means a steel devoid of non-metallic inclusions and free of surface cracks. To do this, the elements of Algoma's Steelmaking and Continuous Casting processes would be refined such that the key characteristics of the semi-finished bloom product would be such that they would satisfy the demands of the rolling mill in making an acceptable finished product. This required that all facets of steelmaking and continuous casting be looked at and improved.

Many improvements were also made in our raw material supply and ironmaking facilities. It is not within the scope of this paper to discuss the many improvements made in those areas. Suffice to say, hot metal quality from our blast furnace has improved in terms of compositional consistency and is the corner stone to having "clean steel."

MELT SHOP PRACTICE IMPROVEMENTS

There have been four major areas of improvement in our melt shop practices:

-- improved B.O.F. turndown performance
-- improvements to the steel transfer process
-- improvements in ladle preparation
-- statistical process control (s.p.c.)

Improved B.O.F. Turndown Performance

Two significant process changes were introduced in the melt shop to promote tighter end point control in temperature and chemistry. One of these changes was the introduction of the lance bubbling equilibrium system (L.B.E.) This system involves the simultaneous blowing of inert gas through permeable refractory bottom plugs to coincide with the normal top oxygen blow of the steelmaking process. A diagram of the L.B.E. system is shown in Fig.1.

The second major change that improved end-point control was the introduction of a static charge model in the melt shop. The static charge model takes measured input and aims, and by energy and mass balancing is able to adjust the blowing parameters to achieve the end point in steelmaking. This sophisticated computerized charge calculation system results in dramatic turndown performance improvement.

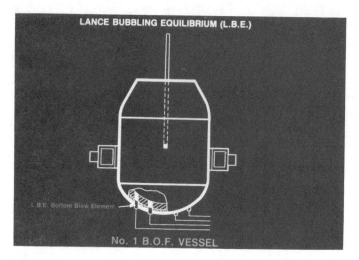

Figure 1

The metallurgical consequences of improved turndown and its impact on chemistry control and cleaner ladle steel can be summarized as follows:

1) Superior phosphorus removal and lower phosphorus levels at turndown.

2) Higher sulphur partition ratios.

3) Less manganese loss to the slag; therefore higher residual manganese at turndown.

4) Fewer reblows.

5) Lower oxygen levels in the steel at tap for a given carbon.

6) Less iron loss to the slag; therefore, higher yield.

7) Lower refractory consumption in vessels and ladles and therefore fewer non-metallic inclusions.

8) More timely delivery of steel to the caster, thereby minimizing low tundish connections.

IMPROVEMENTS IN STEEL TRANSFER PROCESS

A sharp reduction in non-metallic inclusion formation was achieved in a program begun in 1984 to reduce the effects of silica. The changes that were introduced at that time include:

    I)   Installation of ladle preheaters
    II)   Upgrading of Refractory Systems
    III) Introduction of slag-free tapping
    IV)   Utilization of artificial insulating slags

I) Ladle Preheaters

The ladle preheaters and the ladle arrangement for the melt shop can be seen in Fig. 2. The preheaters swivel horizontally and the ladle on a car is pulled towards the furnace. Ladle preheaters were necessary to allow conversion of the ladle refractories from fireclay brick (which contains high levels of silica) to high alumina brick.

Figure 2

II) Upgrading of Refractory Systems

Ladle refractories were replaced with high alumina brick. Campaign life has increased by a factor of three and therefore non-metallic exogenous inclusions have been reduced by a similiar amount from ladle refractory origin. Ladle pre-heating has improved temperature control by eliminating 'cold' ladle effects.

III) Introduction of Slag-Free Tapping

Fig. 3 shows the equipment used in the slag-free tapping process. An engineered slag trap ball has been designed to float at the slag-metalinterface and, when injected at the vortex of the tapping stream, will restrict the amount of slag flowing through the tap hole at the conclusion of the tap. (Fig. 4) This practice, with improved tap-hole maintenance programs which achieved a tighter tapping stream to minimize reoxidation inclusion pick-up, has minimized furnace slag carryover.

During tapping, artificial slags (calcium aluminate) are added to the steel for treatment. At the end of the tap, artificial insulating slags are added to reduce temperature losses. These slags are included when slag on the ladle is measured at Algoma Steel.

Figure 3

Macro-etch tests of blooms done at Algoma have indicated that to
enhance product quality, it is necessary to restrict slag depth
in the ladle to less than 5 inches. Ninety two percent of heats
tapped have less than 3 inches of slag. We currently average 1.9
inches of slag on the ladle. The slag depth measured here
includes artificial slags added for insulation and treatment
(Calcium Aluminate) during the tapping of the steel from the
furnace.

IV) Utilization of Artificial Slags

Furthermore, the development of the slag-free tapping system now
allows for the possibility of insulating the steel ladles with
engineered artificial slags that are low in the oxides of
silicon and iron. The combined effect of reduced slag carryover
is a net reduction of inclusions as indicated in the monitoring
of bloom internal quality. It also allows the treatment of the
steel with calcium.

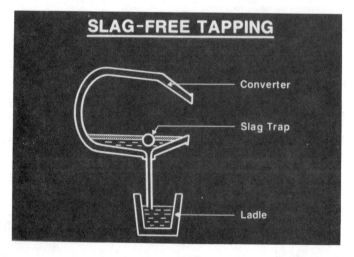

Figure 4

IMPROVEMENTS IN LADLE STEEL PREPARATION

After the steel is tapped out of the furnace into the ladle, it must be prepared for the continuous caster.   Improvements in ladle steel preparation include an argon stir station or ladle treatment station, ladle stir plugs and ladle covers.

During the tap, alloys in bulk are added. The alloys must be trimmed very precisely to fit the required end use. To do this, Algoma has installed what we call a Ladle Treatment Station equipped with wire feeders which we use for trimming (Fig. 5) and a ladle cover that is raised and lowered hydraulically and is designed to draw away fumes generated during stirring and alloy additions.   The wire may be solid as in the case of aluminum, but usually it is a steel shell with a cored center as for carbon, manganese and calcium.   Wire trimming has several important advantages:

1)   It is very reliable.
2)   It is very precise.

Figure 5

Not only must the steel have the correct chemistry, but it must be at the correct temperature and the temperature must be homogeneous throughout the ladle. This is accomplished at Algoma Steel by stirring an inert gas through a porous plug in the bottom of the ladle. The capability to top stir through a lance inserted through a trap door in the cover has also been kept. The cover is raised and lowered hydraulically and is designed to draw away fumes generated during stirring and alloy additions.

Inert gas stirring of the steel is critical in achieving a high degree of steel cleanliness by washing out and floating non-metallics into the artificial slag. Other benefits include:

1)   Improved chemistry control by making the bath more homogeneous.

2)   Improved temperature control.

Good temperature control promotes an optimal balance of equiaxed
and colunar zones in dendritic formation upon solidification in
the casting machine.

The effect of stirring on temperature control during casting is
a net reduction in the difference between start cast temperature
and end cast temperature for each ladle. Improved strand quality
through reduced surface and internal cracking as well as the
degree of centerline segregation and porosity can be directly
related to cast structure which is highly dependent on
temperature as well as chemistry control. Improved temperature
control has also been enhanced through the use of ladle covers.
Fig. 6 is a picture of a ladle cover in place during casting.

Figure 6

Statistical Process Control Techniques

Improved quality has also being enhanced through the use of
Statistical Process Control techniques. An S.P.C. team working
on turndown performance has been able to define a process
capability and through the tools of S.P.C. are now working to
reduce the standard deviation of the various input parameters so
as to achieve an improved process capability. This team is also
the nucleus for implementing, to the shop floor, the on-line
monitoring of various product and quality characteristics and
process elements through the use of control charts. Operating
personnel at all levels are being subjected to more
accountability for quality in each elemental area.These are the
key process elements that are being monitored:

At the BOF -- turn down temperature
             delivery temperature
             chemistry
             furnace slag
At the bloom caster --
             surface quality
             internal cracking
             internal cleanliness
             volume control

PRACTICE CHANGES AT THE BLOOM CASTER TO ACHIEVE "CLEAN
STEEL"

1) Upgrading of Tundish Refractories

Magnesite boards have replaced silica boards to improve quality.
More recently high density pre-heatable magnesite boards with a
non-organic binder have replaced magnesite boards made with an
organic binder.   These boards eliminate hydrogen evolution from
the organic binder, improve temperature control. at start-up (by
preheating), and are more durable.

2) Submerged Casting

Fig. 7 is a schematic depicting a submerged casting sequence of
a fully shrouded system for the casting of aluminum-deoxidized
steel as done at Algoma Steel.   This system gives a completely
enclosed pouring stream from ladle right through to mould.
Nowhere is there attack from oxygen.   Nowhere is there a place
for dirt to form. This significant improvement made in 1984 had
the following benefits:

a) Improved protection from oxygen attack; thereby reducing
reoxidation product.

b) Mould powder lubricant replacing oil lubricant reduced mould
   joint defects and bloom cracking. Mould powder has the added
   benefits of acting as an insulator on top the steel in the
   mould and prevents crusting from occurring. The mould powder
   also acts as a fluxer for the removal of any inclusions
   floating up.

c) Aluminum-deoxidation of steel rather than silicon results in
   reduced pin holes and also reduced size and number of silicate
   inclusions.

d) Improved strength through the use of grain refiners.

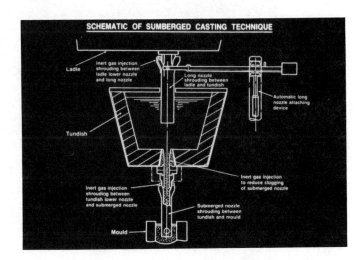

Figure 7

3) Tundish Flow Control with Dams and Weirs

Flow control improvements were made in the tundish. This is an
area that has proved to be extremely important. Fig. 8 depicts
the flow of steel from a submerged entry nozzle in the tundish
with no flow control.   Notice that the path of steel is much
shorter for the two inside strands than it is for the outside
strands. Notice too, that the flow is essentially downward and
lateral.   Notice the large areas above the flow lines which are
essentially dead zones. Any inclusion in the stream of steel
will be swept down the tundish nozzles into the mould.

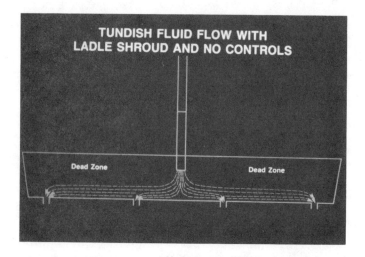

Figure 8

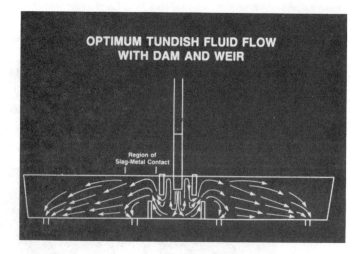

Figure 9

4) Mould Fluid Flow Improvements

Water modelling at the University of Toronto was done to determine the optimum shroud design that would give the best flow pattern within the concast mould. Fig. 10 and Fig. 11 depict the basic configurations investigated.

The initial work done suggested, that the configuration in Figure 10 (4 ports) was the superior design. Its main advantages were less penetration of the stream as the stream was directed outwards and upwards which would facilitate flotation of inclusions and allow them to be fluxed into the covering mould powder. Its big disadvantage was that inert gas bubbles injected into the stream to prevent clogging were also directed at the side of the mould where they tended to get caught in the freezing front causing sub-surface porosity and eventually seams in the finished product.

The design in Fig. 11 had the disadvantage of too much penetration of the steel making flotation of any inclusions difficult. It was found however, that the penetration could be reduced and flotation enhanced by increasing the bubbling rate of the inert gas. The inert gas could escape through the shroud mould powder interface. It was also found that the increased inert gas bubbling was necessary as it created a gentle rolling action on the surface that prevented the formation of surface crusts and hence eliminated all subsurface porosity from gas entrapment. This proved to be the superior design for production and is the design that was ultimately chosen.

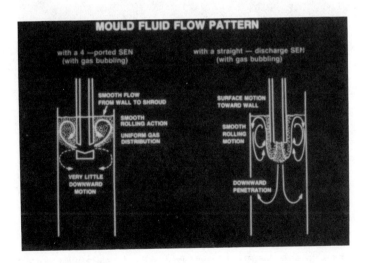

Figure 10   Figure 11

5) Automatic Mould Level Control with Tundish Slide Gates

Automatic mould level control with Tundish Slide Gates was
installed in February of 1986.   Fig.12 is a schematic of the
slide gate system used at Algoma Steel.   This system
incorporates the Berthold Mould Level detection equipment which
employs a low strength radioactive source.   The electronic level
control equipment takes the signal from the level indication
equipment and using its own built-in logic, automatically
controls the operation of the hydraullically-riven slide gates,
which physically regulate the rate of steel flow into each mould.
Algoma Steel believes the slide-gate system to be the best
available system, being more durable, allowing more positive
shut-off control thereby being safer to operate, and being more
precise in its throttling control.

This system significantly improves the surface quality of the
powder lubricated cast product and yet maintains the high
internal quality achieved with submerged casting. The slide-gate
system is however, totally unforgiving when casting aluminum
killed steels.   To cast with slide-gates, correct calcium
treatment of the steel is mandatory.   From a quality point of
view, slide gates can be viewed as almost self-policing.   That is
to say, non-clean steel cannot be cast through the tundish slide
gates.   They clog almost immediately, if steel preparation is
inadequate.

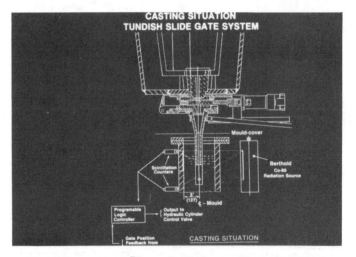

Figure 12

6) Enlarged Tundishes

All tundishes have been converted at the bloom caster in 1986
from a depth of 22 to 30 inches.   This conversion allows for
improved oxide removal.   By increasing residence times through
tundish flow control and increasing the depth of the tundish, the
theoretical size of the smallest inclusion that can be floated
out is decreased.   Only the smallest of inclusions can pass
through the pouring system.

7) Bloom Test Cutting Station and Laboratory Requirements

A bloom cutting station does not sound like something significant for direct charging. But it is. If you plan to direct charge, it is advisable that you have a test cutting station that can quickly cut tests for sulphur prints or macro-etches before the heat is shipped out. With direct charging, sulphur print results will come available after the product is rolled. Direct charging will also place new strains on the chemistry laboratory, as final chemistry results (taken in the tundish at 10, 30, 50 minutes after start cast at Algoma) must be known before the heat is shipped out.

Off analysis heats, are particulary troublesome, when trying to make hot connection with a rolling mill. For this, it is imperative that you know where your "end of heat" is within your cutting lengths. At Algoma, "end of heat" and transition zones were measured using nickel tracers in alternate heats in the same series. Because of these tracer trials, "off-analysis" bars held back from rolling are minimized.

PLANNED IMPROVEMENTS

While direct charging goes on at Algoma, further improvements to the following facilities are planned and currently are being engineered to vastly improve steel refining in No.1 Steelmaking:

        --dephosphorization and reladling
        --ladle arc-reheating
        --vacuum degassing
        --powder injection

The impact of quality that will follow upon completion of these projects include:

        --superior phosphorus removal through reladling and
          lower furnace tap temperatures.
        --enhanced sulphur removal and therefore ultra-low
          desulphurization capability
        --enhanced control of type and shape of non-metallic
          inclusions (oxides and sulphides).
        --removal of gases such as hydrogen and oxygen
        --improved time and temperature delivery to the
        caster and therefore ideal casting conditions.

This concludes the review of the improvements made in the No.1 Steelmaking complex at Algoma Steel. The question now is does it work? The answer, when direct charging was first attempted was "yes", then as other problems were solved, new problems surfaced and the answer was "no" as we struggled on a learning curve. The answer today is "yes", direct charging of blooms works very well.

PRODUCTION RESULTS

One of three main products from the bloom caster is rails.
The bar chart in Fig. (13) shows the index of first class rails
produced for all concast rails made in the years 1984, 1986,
1987, and 1988. 1984 is the last full year where all rail blooms
were inspected. 1985 was left out because some rail blooms were
inspected and some were direct charged. It can be seen there was
a drop off in performance in 1986. The reason for this drop off
had to do with our submerged entry nozzle design which was
switched from a "ported" to a "straight through" submerged nozzle
and changed from argon to nitrogen for bubbling medium.

Once the submerged entry nozzle design was sorted out,
performance levels surpassed 1984 levels. It should be pointed
out that comparing 1984 performance levels with 1988 is not
really fair to the 1988 performance and does not truly depict the
improved quality.

In 1984, cast rails being produced through the caster were 38
foot lengths, were carbon grades and were inspected visually.
After 1985, cast rails are up to 80 feet in length, are carbon,
intermediate alloy or chrome alloy grades and are inspected
"online" ultra-sonically and with magnetic particles and black
light. This came about due to a major upgrade in our rail
finishing facilities.

In conclusion, due to many changes made in the primary end,
direct charging of rail blooms is now possible and being done.
The quality of the finished product has significantly improved.

Rail Inspection -- All Concast Rails

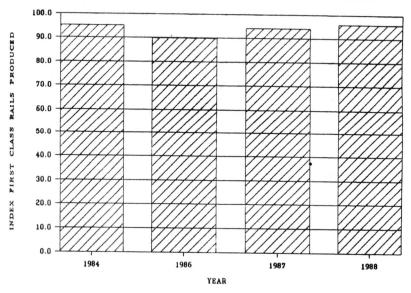

Figure 13

DIRECT ROLLING AT NUCOR STEEL-NEBRASKA

J. Schoen

Nucor Steel, Norfolk, Nebraska

ABSTRACT

This presentation discusses direct rolling at Nucor Steel of Norfolk, Nebraska. The first area discussed is why Nucor chose direct rolling for the Nebraska operation. The subjects of this section include product size, product grade, and billets which are not able to be direct rolled. Secondly, experience with induction heating is discussed by means of comparing it with the conventional gas fired reheating furnace. Topics discussed here include heating efficiency, personnel requirements, reduced scrap by controlling crop losses, and scale thicknesses.

KEY WORDS

Metering nozzles, heating efficiency, personnel requirements, crop loss, scale thickness, downtime.

INTRODUCTION

Nucor Steel of Norfolk, Nebraska is one of the Nucor Corporation's four steel making facilities. Of these four plants, two are able to direct roll by means of induction heating as the billet reheating source. The plants of Darlington, South Carolina and Norfolk, Nebraska each have one rolling mill which employs induction heating. The subject of this presentation concerns the casting and rolling operations of the Nucor – Norfolk facility.

In order to exhibit more fully our experience with induction heating, we will also present some information about our Norfolk rolling mill, Nucor – Norfolk I, which uses the conventional gas fired reheating furnace.

HISTORY – WHY WE USE INDUCTION HEATING

Nucor – Norfolk I (NNI) has been melting since September 1973 and rolling since February 1974. During the six year period from 1973 to the start-up of the second mill in October 1979, a great burden was placed upon the rolling operation with regard to the wide range of different sizes and shapes of the finished product. The products produced at that time are listed in Fig. 1

Angles	1 x 1 x 1/8, 3/16, 1/4
	1 1/4 x 1 1/4 x 1/8, 3/16, 1/4
	1 1/2 x 1 1/2 x 1/8, 3/16, 1/4
	1 3/4 x 1 3/4 x 1/8, 1/16, 1/4
	2 x 2 x 1/8, 3/16, 1/4, 5/16, 3/8
	2 1/2 x 2 1/2 x 3/16, 1/4, 5/16, 3/8
Structural Angles	3 x 3 x 3/16, 1/4, 5/16, 3/8
	3 1/2 x 3 1/2 x 1/4, 5/16, 3/8
	4 x 4 x 1/4, 5/16, 3/8, 7/16, 1/2
Unequal Leg	2 1/2 x 2 x 3/16, 1/4, 5/16, 3/8
	3 x 2 x 3/16, 1/4, 5/16, 3/8
	3 x 2 1/2 x 1/4, 5/16, 3/8
	4 x 3 x 1/4, 5/16, 3/8, 7/16
Channels	3 x 4, 1, 5, 0
	4 x 5, 4
Flat Bar	1/4 x 2, 2 1/2, 3, 3 1/2, 4, 4 1/2, 5
	5/16 x 2, 2 1/2, 3, 3 1/2, 4, 4 1/2, 5
	3/8 x 2, 2 1/2, 3, 3 1/2, 4, 4 1/2, 5
	7/16 x 2, 2 1/2, 3, 3 1/2, 4, 4 1/2, 5
	1/2 x 2, 2 1/2, 3, 3 1/2, 4, 4 1/2, 5, 6
	3/4 x 6, 1 x 6
Rounds	1/2, 9/16, 19/32, 5/8, 21/32, 11/16, 23/32,
	3/4, 25/32, 13/16, 27/32, 7/8, 15/16, 1,
	1 1/8, 1 1/4, 1 3/8, 1 1/2

Fig 1. Products produced by NNI from 1973 to 1979.

After considering the projected increased demands of Nucor's product, the decision was made to build a second mill at the Norfolk location. Nucor - Norfolk's desire to improve quality and productivity in the existing mill without major equipment changes would shape the parameters of the new mill, Nucor - Norfolk II (NNII), which was to start-up in October of 1979. The idea of that time was to make the larger sizes of all shapes in NNI and the smaller sizes in NNII. It was also desired to produce all sizes and grades of rounds in NNI and primarily A-36 and A-36 modified channels, angles, and flats in NNII. This would mean less downtime in both mills since the sizes and shapes produced in each mill were of a narrower range than the original variety of sizes of NNI. Nucor - Norfolk was also shifting its capabilities to be more customer oriented with respect to what the finished product would be. We wanted to sell what the customer wanted instead of simply what we wanted to make. From these considerations, the annual capacity (projected at 160,000 tons), the product to be made, and the billet size were arrived at. These factors would permit the use of the more efficient direct rolling with induction heating. But induction heating in turn, would influence both the NNII caster and the NNI rolling mill as the discussion below points out.

Since the product size for the new mill would be smaller, the billet size was decided to be 4 3/8 inches square instead of the 5 1/4 inches square billet cast in the NNI operation. The product size was arrived at after a marketing study was made in order to determine the size range needed to operate a direct rolling mill profitably and quality consciously. The new caster, designed and built by Nucor as in the old mill, would have two strands. Since the rolling mill was to use induction heating as the billet reheating source, more than two strands would cause too much of a temperature loss before some of the billets could be placed into the heaters. (For our operation the heaters have only the capability to reheat the surface from 1800°F to 2200°F. On some of the smallest sizes, only one casting strand can be used because of the rolling speed and this limited capability of the induction heaters. Varying the size of the caster metering nozzles has helped but not solved this problem of matching the caster's ability to the rolling mill's and thus the induction heaters). This limitation of heating ability brought up the question of what to do with billets which are too cold to reheat after being delayed by problems associated with the rolling mill or the heaters themselves. This was probably one of the most decisive factors for Nucor's implementation of induction heaters for direct rolling. The answer to this problem would be associated with NNI.

Between 95% and 100% of NNII's products would be various sizes of A-36 and A-36 modified. Since NNII's finished products would be of similar cross-sectional area, the chemistry aims of each heat of A-36 and A-36 modified would be close to one another. From this, the cold billets of these two grades could be stacked in separate areas of NNII. One area for A-36 and one for A-36 modified. Since the gas reheat was already employed at NNI, these cold NNII billets could be moved, by rail, to the NNI reheat furnace and rolling mill. This would minimize the amount of scrap generated simply because the billets were to cold to reheat. The numbers of tons of A-36 and A-36 modified transported to NNI for the past four years are shown in Fig. 2.

As mentioned earlier, up to 5% per year of the NNII product would be grades other than A-36 and A-36 modified. For reasons of quality, these discarded billets would be cut into scrap. But over the years, we have found that this could be minimized by scheduling these grades to be cast several heats after the appropriate size change in the rolling mill. This way, adjustments can be made early without delaying the non-A-36 grades.

Overall, direct rolling with induction heating was more than a decision to have it simply for heating efficiency. It was a decision based on if we could use induction heating with the product we wanted to make and how cold billets which were not rolled could be salvaged.

EXPERIENCE WITH DIRECT ROLLING

When comparing the Nucor Corporation's mills of NNII's rated annual capacity, NNII is the most cost efficient mill of the corporation. This is due to a number of reasons which we will try to illustrate by means of a comparison between our own gas fired reheat furnace and our induction heaters. Operational parameters of Nucor's Darlington mill were not available for the writing of this report.

The most obvious reason for NNII's cost efficiency is the use of the billets' residual heat from melting and the fact that energy is used to heat only the billet and is used only while the billet is passing through the heaters. There are four Westinghouse, 3600 kw, 1000 Hz induction heating units employed in series with one another and with the rolling mill and caster. See Fig. 3.

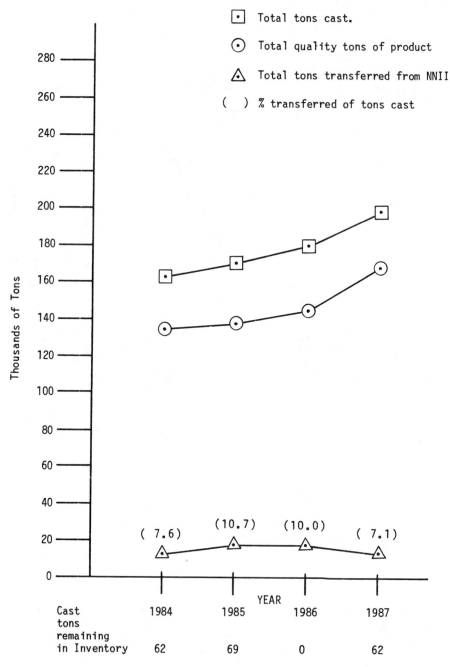

Fig. 2   NNII casting and rolling performance 1984 to 1987.

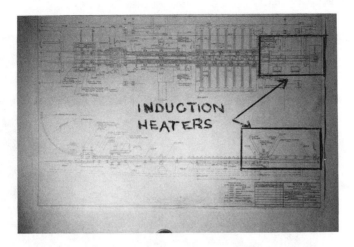

Fig. 3   Layout of NNII direct rolling mill.

The lower the frequency the greater the depth of penetration. NNII aims for about 1" of heating penetration. The 1987 cost of billet reheating at NNII was about $2.40 per ton. This comes from the average utility rate $0.04 per kilowatt hour and an average power consumption rate of 60 kilowatt hours per ton of steel heated.

Fig. 4. Induction heaters in operation

The NNI gas reheat has all of the well know disadvantages concerning energy efficiency. These include lost heat in the flue gas, refractories, and the furnace shell. The average 1987 cost per ton of steel heated in NNI was about $4.64 per ton. This figure is based on the average natural gas consumption rate of 1.6×10^6 BTU per ton and the average utility rate of $2.90 per million BTU. This is the

average cost of both direct charged and cold charged heats. In 1987, 45.2% of NNI
heats were direct charged.

Fig. 5 NNI gas fired reheat furnace.

Another cost saving advantage of our direct rolling operation is the number of
people required for the removal, accounting, and reheating of billets. Table 1
shows the breakdown of personnel required for the two mills.

Table 1 Comparison of Required Personnel

Job	NNI	NNII
Caster Shear Pulpit	1	1
Billet Charging Pulpit	1	
Billet Removal Crane	1	
Billet Yard Accounting and Coordination	1	1
Billet Transferring (NNII to NNI)	0	
TOTAL	4	2

As the table points out, only half the people are required for the direct rolling
heating operation compared to the gas fire reheat operation.

Over the years we have found another advantage of direct rolling which is often not
pointed out in induction heating writings. It is the advantage of being able to
adjust the billet length to control crop loss while casting and rolling the same
heat. Even though billet lengths are precalculated for each rolling mill size and
length, the lower portion of the range of a bar tolerance can lead to lost material
in the form of shorts or crop loss. By adding only inches to the billet length
while the heat is being cast, this loss can be practically eliminated. The holding

time in a gas fired reheating furnace does not permit this quick adjustment.

The typical scale thickness from our induction heating operation varies from .017" to .027" with an average thickness of .022". Those from the gas fired operation vary from .026" to .045" with an average thickness of .034". This is over 50% more scale produced from the gas fired furnace. These scale thicknesses were measured with a micrometer from randomly selected A-36 billets which were direct charged from the caster with a minimum time in the furnace and direct rolled A-36 billets.

Other advantages concerning the induction heaters are the absence of pollutants, a cleaner work place, and reduced downtime due to the induction heaters themselves. When an induction heater fails it can be completely replaced with another unit usually within one hour. Relining the gas fired furnace has been the major NNI reheat downtime we have observed over the past few years. Complete relines are scheduled during maintenance shutdowns and so are never a threat to production. Occasionally when minor refractory repairs must occur at times other than scheduled shutdowns, the cooling period alone, at best, takes 24 hours.

Up to this point only the advantages of direct rolling have been discussed. One disadvantage is that the scheduled caster downtime has to be coordinated with the scheduled rolling mill downtime. If tundish and mold changes are not coordinated with rolling mill changes, the caster is forced to put billets on the ground or the rolling mill is without steel. In 1987 the NNII rolling mill was without steel 5% of the time. This nonproductive time was due to breakouts, unscheduled mold and tundish changes, and melting problems.

As mentioned previously, the NNII melt shop and rolling mill produces primarily merchant quality bars. The caster employing induction mold stirring, pollard shrouds, and nuclear radiation mold level controllers produces an excellent quality merchant bar that is satisfying our customers very well. And direct rolling with induction heating has served this operation well. Induction heating in our process enables us to heat billets from a wide range of temperatures below the process rolling temperature to a temperature within the process rolling temperature range. Varying the sizes of the metering nozzles has helped to control the casting speeds and thus the entry temperatures to the induction heaters. Nozzles can be changed and the tundish back in production within three minutes.

Over the years we've adapted induction heating to suit the capabilities of the mill. This success of direct rolling at NNII has enabled NNI to expand its market into the area of special bar quality rounds for cold heading and cold drawing applications with the grades ranging from low to high carbon to resulphurized to mild alloy steels. NNI now employs an eccentric bottom tapping furnace, a ladle metallurgy station, mold induction stirring, pollard shrouds used in conjunction with canopy shrouds, and nuclear radiation mold level controllers. A new three strand caster and new gas fired reheat furnace will be in operation by December of 1988. A new rod mill with Nucool cooling and Easy Draw Cooling has just been added to serve customers who process rounds down to 7/32" diameter. All of these improvements have been and will be made to better serve our customers with a higher quality product. With this, Nucor-Nebraska's product has evolved into that shown in Fig. 6.

NNI

```
Angles 2 1/2 x 2 1/2 x 3/16, 1/4, 3/8
       3 x 3 x 3/16 1/4, 5/16, 3/8 1/2
       3 1/2 x 3 1/2 x 1/4, 5/16, 3/8
       4 x 4 x 1/4, 5/16, 3/8, 1/2
```

```
       Channels 3 x 4. 1, 5.0
                4 x 5.4, 7.25

   Bar Channels 1 1/2 x 9/16 x 3/16
                1 1/2 x 1 1/2 x 1/8
                2 x 1 x 1/8, 3/16
                2 x 9/16 x 3/16
                2 x 1/2 x 1/8
                2 x 5/8 x 1/4

       Flat Bar 3/16 x 4
                1/4 x 4, 4 1/2, 5
                5/16 x 5, 4 1/2, 5
                3/8 x 4, 4 1/2, 5
                1/2 x 4, 4 1/2, 5
                5/8 x 2, 2 1/2, 3, 3 1/2, 4, 5
                3/4 x 2, 2 1/2, 3, 3 1/2, 4, 5
                1 x 2, 2 1/2, 3, 3 1/4, 4
                1/4 x 6, 5/16 x 6, 3/8 x 6
                1/2 x 6, 5/8 x 6, 3/4 x 6
                1 x 6, 1 x 5

        Squares 1/2 5/8, 3/4, 7/8, 1, 1 1/4, 1 1/2


         Rounds 1/2, 9/16, 19/32, 5/8, 21/32, 11/16, 23/32
                3/4, 25/32, 13/16, 27/32, 7/8, 15/16, 1
                1 1/8, 1 1/4, 1 3/8, 1 1/2, 1 5/8, 1 3/4, 2
                Rod Mill sizes will be 7/32 to 5/8

                          NNII

         Angles 1 x 1 x 1/8, 3/16, 1/4
                1 1/4 x 1 1/4 x 1/8, 3/16, 1/4
                1 1/2 x 1 1/2 x 1/8, 3/16, 1/4
                2 x 2 x 1/8, 3/16, 1/4, 3/8

Unequal Leg Angles 2 x 1 1/2 x 1/8, 3/16, 1/4
                2 1/2 x 2 x 3/16, 1/4, 5/16, 3/8

       Flat Bar 3/16 x 1 1/2, 2, 3, 3 1/2
                1/4 x 1, 1 1/4, 1 1/2, 2, 3, 3 1/2
                5/16 x 1, 1 1/4, 1 1/2, 2, 3, 3 1/2
                3/8 x 1, 1 1/4, 1 1/2, 2, 2 1/2, 3, 3 1/2
                1/2 x 1, 1 1/4, 1 1/2, 2, 2 1/2, 3, 3 1/2
```

Fig. 6 Current Products of Nucor Steel

As NNI's capabilities are moved away from merchant quality, the billet size of NNII may be increased to 5 1/4". NNII will again lighten the burden of the original mill — probably with many of the larger sizes currently produced in NNI. With this larger billet, we are confident that direct rolling will continue to be a valuable asset to both Nucor Nebraska's quality and cost efficiency. For Nucor-Nebraska, direct rolling has been and will continue to be an essential part of our success.

SUMMARY

Direct Rolling is used at Nucor-Nebraska for these Reasons

1. Heating efficiency.

2. The range of product sizes and shapes provides a low amount of change over time.

3. The availability of a gas fired reheat furnace.

4. Most of the product is of the same chemistry.

It is not simply a matter of installing induction heating for heating efficiency but more a question of if one can use induction heating.

Advantages of Direct Rolling

1. The heating efficiency of induction heaters.

2. Reduction in the number of people required.

3. Good control over crop loss.

4. Less scale is generated.

5. Cleaner air, a cleaner work place and reduced downtime due to refractory maintenance.

Disadvantages of Direct Rolling

1. The downtime at the caster must be coordinated with the downtime at the rolling mill.

2. The capabilities of the caster need to be closely matched with the capabilities of the rolling mill.

ACKNOWLEDGMENT

Thanks are due to the following Nucor Steel – Nebraska personnel who have contributed advice and support for this paper. John Doherty – Vice President, Nucor Corporation, General Manager Nucor Steel – Nebraska, John Troutman – Melt-Cast Manager, Gene Tyson – Sales Manager, Leroy Stebbing – Special Projects Engineer, Dave Mauch – Accountant, and Doug Cheney – Safety Coordinator.

DIRECT ROLLING AND DIRECT CHARGING OF STRAND CAST BILLETS

Dr. Walter Maschlanka and Dr. Siegfried Henders

Deutsche Voest Alpine Industrieanlagenbau GMBH vorm. Korf Engineering GMBH

1. Introduction

In the processing route comprising of melt shop, continuous casting plant and rolling mill, the development stage of the two interconnecting areas is rather different. While up till now particular attention has been paid to the interrelation between the steel melt shop and continuous casting plant, the second important interconnection, namely between the continuous casting plant and rolling mill, is gaining in significance.

In continuous casting plants, usually located directly at the melt shop, the entire liquid steel production can be cast into bars by coordinating the parameters determining the throughput. In order to make sequence casting possible, a ladle furnace undertakes, in many instances, a certain buffer function. These considerations led to a ladle furnace being constructed at Marienhütte as early as 1983.

The second inter-connection area, namely between the casting plant and rolling mill, is being improved gradually whereby, in the first stage, hot charging of the cast product into the pusher furnace is being tested. As can be gathered from Table 1, energy expenditure is reduced by approximately 0.6-0.8 GJ/t at a drawing temperature of 1,150°C, when the billets are hot charged at 600°C into the pusher furnace.

This type of inter-connection is already being tested at many locations, however, the charging temperature, in particular with small-sized continuous casting, is on average, below 600°C. The next stage of inter-connection between continuous casting and the rolling mill is direct rolling of the cast billet.

This development of the direct linkage between the casting plant and rolling mill is shown in Figure 1. This was the reason for the decision to install a Rocast® machine (Fig.2) at the Marienhütte plant, at Graz, Austria, where the following improvements were made:

- Cutting down on energy by utilizing the heat contained in the casting billet for subsequent hot forging
- Greater productivity and yield
- Fewer manpower requirements
- Elimination and reduction of billet storage

Both the spatial requirements for favorable material flow as well as the co-ordination of the process parameters of the two plants are prerequisites for the successful combination of continuous and direct rolling.

Favourable prerequisites for such an inter-connection were given at Marienhütte due to the limited number of steel qualities and favourable lot sizes of the various product dimensions.

2. Integration of a Rocast® Plant at Marienhütte

In planning the Rocast® plant as an intermediary link between the steel and rolling mill, the following conditions were to be met:

- In the construction phase, normal steel works operation was to be maintained without any fundamental disturbances
- The Rotary continuous caster was to be sited as close as possible to the rolling mill, so that both in-line operation as well as continuous billet supply from the furnace would be possible.
- Long billets were to be alternatively produced for sale as semi-finished products.

Figure 3 shows the ground plan of the steel works. Scrap (as a basis) is melted in an arc furnace with an 18MVA transformer capacity and an average tap weight of 27t. The molten steel is subsequently treated in a ladle furnace with a maximum transformer capacity of 4.5 MVA.

The ladle treated in this manner is placed on a rail mounted cross-transport car by the furnace hall's crane, transported to the rolling mill hall and brought directly into the casting position from where the steel is fed to the casting mould from the ladle via a tundish. Start up of casting is carried out with a dummy bar similar to a conventional plant.

The 165/155mm × 128mm continuously cast billet is rolled down after the shears to 120mm × 120mm in a vertical and horizontal stand configuration. Depending upon the production programme, either 10m long billets are returned to a cooling bed via a parallel roller conveyor, or short billets 3.50m long are fed to the trio-stand of the existing rolling mill via a rapid cross transfer.

The plant is designed for sequence casting. Due to the fact that the crane of the rolling mill hall has only the carrying capacity for an empty ladle, the transport car has been placed high enough so that the casting position can be reached directly.

3. Operational Experience and Adaptation Measures

In the first trials to roll billets cast in the casting wheel plant directly by by-passing the pusher furnace, it was discovered that the capacity of the rolling mill was reduced by about 20% due to malfunctions at the trio-stand. The cause for this was a billet temperature lower than that from the usual pusher furnace procedure when entering the trio-stand, leading to the partial overloading of the drive. The surface temperatures measured under the specified conditions were approximately 1,080°C. To raise this temperature the following measures were considered.

The casting speed naturally has the greatest influence on the surface temperature. Here, in measuring the casting speed and casting size, a compromise had to be accepted. On the other hand, to reach a high number of casts in sequence, the casting production should not be more than 20t/h, which is the capacity of the melt and the capacity of the rolling mill, despite the fact that the Rocast® plant would have been in the position to work at a considerably higher casting rate, which would probably have been sufficient to attain a high enough billet temperature when entering the trio. On the other hand, reducing the casting size was thought not possible, since commercial billets of 120mm square were to be produced.

Bearing these requirements in mind an average pouring rate of 2.25m/min. was arrived at, corresponding to a casting performance of 20t/h.

For this reason other measures had to be adopted to eliminate the difficulties resulting from the heat loss. To begin with, the quantity of spray water from the secondary cooling system was reduced by approximately 70-75%. Other measures were to partially insulate the transport path between billet discharge from the Rocast® machine and, additionally, the billet entry in the rolling mill's trio, and to increase the transport velocity. The roller table behind the spray chamber and pull-out driver were insulated with hydraulically hinged hoods. The existing slow-operating shears were exchanged for pendulum shears, whereby it was possible to insulate the roller conveyor up to the billet entry in the V-stand. The effect of these insulation measures was initially determined by a computer model. As can be seen from the isotherms shown in Figure 4, it was expected that these temperature conditions could be considerably improved by these measures. These computer figures were later confirmed by measurements.

4. Operating Results

As already mentioned, a number of adaption measures were necessary subsequent to the Rocast® plant being put into operation, so that it could not be expected that the advantages envisaged of coupling the casting plant directly with the rolling mill would be attained in the shortest time.

Before the above-mentioned modifications in the transport system between the casting wheel plant and rolling mill were completed, the Rocast® plant was only used for producing billets, which were either stored for internal consumption or were intended for sale. Despite the fact that staff experienced in continuous casting were available, a certain learning process was necessary, particularly with regard to maintenance and operational organization. For this reason the most frequent causes of malfunctions, i.e. 75.4% of the total, were assigned to this sphere.

The rolling mill was initially supplied with billets preheated in the pusher furnace. In March 1987 the transport system and the insulated section were completed. After this date billets were rolled only directly from the Rocast® plant. Apart from an energy economization of approximately 1.7 GJ/t the other savings, such as in storage, reductions in staff and productivity, had a favourable effect on the overall result. Thus, by way of example, the yield could be increased by approximately 1.2% through less scale being incurred.

Manpower requirements were halved by the pusher furnace operation and the work in the billet storage area was eliminated. Some 250,000 tons of various steel qualities have been produced with the casting wheel up to the present day.

The billet's transport time after discharge from the H-stand prior to entry to the trio was originally over two minutes. Since the billet's surface temperature drops by about 1°C/min when being transported in the air, it was necessary to speed up the transport. Incorporating a rapid transfer skid enabled this time to be reduced to about 40 sec., so that, as the sum of these measures, the entry temperature of the billet into the trio was increased by almost 100°C. Figures 5 and 6 show the course of the surface temperatures resulting from the optimization measures during the procedures of casting, transporting and rolling as a time factor.

On the trio itself, the old friction bearings were replaced by rolling bearings in order to avoid drive overloading, should the specified rolling temperature of 1,150°C not be attained in some cases.

Another measure for ensuring smooth in-line rolling was by reducing the inclination of the side flanks of the trapezoid in the mould. The original (much marked) flank inclination of 6° led to the billets being more or less "ski formed", caused by more material being formed on the base surface in the V-calibre than on the narrow trapezoid side. Furthermore it was observed that varying degrees of stretching of the rolled material over the length of the inclined trapezoid side led to fine surface cracks in this area when high Cu contents were present in the steel. This phenomenon was all the more marked with greater inclination of these trapezoid sides. For this reason the flank inclination was gradually reduced from 6° to approximately 2.3°. The above-described difficulties could be eliminated in this manner without any other drawbacks being incurred as a result. In this way, an angle of 1° only on the side flanks will be tested shortly.

About 10 melting charges are cast in sequence on average today. This is brought about both by the tundish nozzle service time in the continuous casting plant as well as by the cycle times of the individual units. As can be seen from Figure 7, the average tap to tap-time of the electric furnace amounts to 92 min., while the casting time for one melting charge only comes to about 75 min. This difference of 17 minutes must be made up by various dwell times in the ladle furnace. When the sequence commences, a sufficiently large buffer in liquid steel must be created, which is reduced in the course of the sequence until finally, only about 20 min. is available for ladle treatment.

Although the main emphasis in production is Tempcore quality in a weldable reinforced steel, to a lesser degree the following grades were also cast and rolled:

- Low carbon steels, such as St 37
- Naturally hard reinforced steel with medium C content
- Case hardened steels, such as 16 MnCr 5
- Special hard qualities, such as 46 Mn 7

For reasons of market policy, however, Marienhütte's production programme has in the meantime been switched over to such an extent that only a few steel grades are being produced.

An important point in linking a casting plant and rolling mill is maintaining certain parameters for producing finished products of perfect quality. Basically, those defects which may occur with cast billets produced from a Rocast® machine may also occur with conventional continuous casting with the exception of:

- Transverse-cracks caused by oscillation marks
- Cold-welding points
- Surface roughness through heavy oscillation marks.

As extensive investigations of the billet material have shown, the same prerequisites are valid for producing good quality material as is the case with conventional continuous casting.

Due to the non-oscillating casting mould the billets coming from the Rocast® have a completely smooth surface. Due to the long guide in the revolving mould, defects in the billet geometry are hardly to be observed, since uniform cooling of the casting mould is ensured and secondary cooling does not occur for practical purposes.

Formation of burrs on the strip side that has always been suspected only occurs in very rare cases and is, in the final analysis, a question of maintenance.

Summarizing, it can be said that with specific quality parameters being maintained, such as chemical analysis, temperatures, cooling, etc., and with the plant being in mechanically perfect condition, a high-grade quality product can be produced, which is in every way on a par with conventionally manufactured material.

TEMPERATURE OF BILLETS °C	ROLLING TEMPERATURE		
	1150°C	1050°C	950°C
	Q_{tot}(GJ/t)	Q_{tot}(GJ/t)	Q_{tot}(GJ/t)
ROOM TEMPERATURE	1.30 to 2.00	1.20 to 1.80	1.10 to 1.60
300	1.03 to 1.87	0.86 to 1.44	0.75 to 1.25
600	0.69 to 1.20	0.60 to 0.90	0.45 to 0.70
800	0.50 to 0.80	0.45 to 0.70	0.30 to 0.50

Table 1   Energy consumption Q_{tot} for reheating of billets

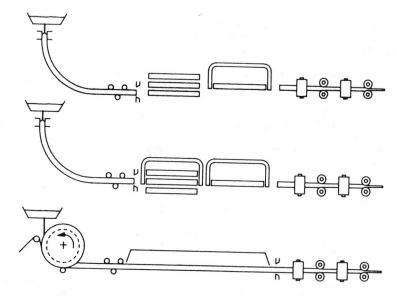

Fig.1   Linkage between cast shop and rolling mill.

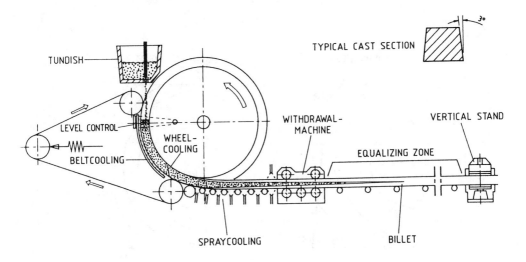

Fig.2   Arrangement of rotary caster.

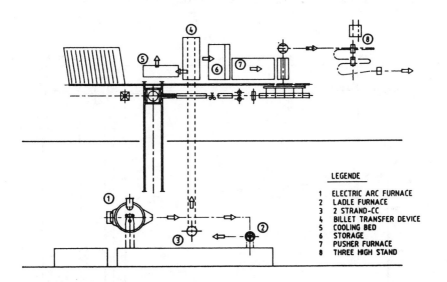

Fig.3   Material flow of Marienhüette 2 strand-CC and reheating of billets.

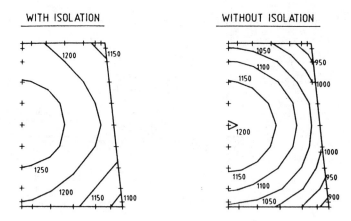

Fig.4   Average surface temperature of billets in V-stand.

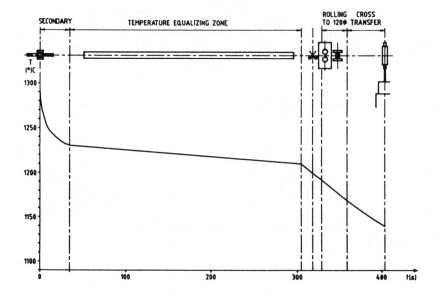

Fig. 5   Billet surface temperatures at Marienhüette plant.

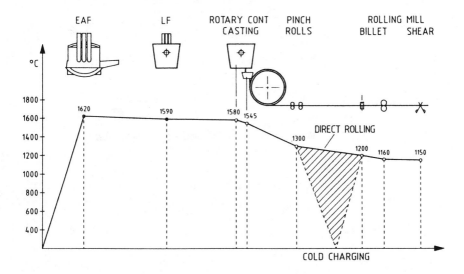

Fig.6   Temperature curve from liquid steel to the rolling mill.

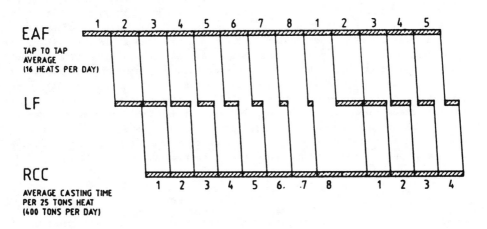

Fig.7   Sequence casting at Marienhüette.

DIRECT HOT CHARGING OF CC BLOOMS OF SPECIALTY
STEELS FOR ROLLING

G. Battan, R. Schlatter and G. Marguerettaz

DeltaCogne S.p.A., 11100 Aosta, Italy

ABSTRACT

The concept of hot charging continuously cast blooms of a wide range of specialty
steels and rolling to intermediate billet sizes was put in operation in 1977. The
original layout and the continued progressive evolution of this casting-reheating-
in line rolling system are described in relation to operational and processing aspects and in terms of product quality for construction and stainless steels.
Recent improvements made in continuous bloom casting, in liquid metal treatment and
particularly in thermal handling of the blooms between caster and reheating furnace
have resulted in a very satisfactory operation system for those difficult to process
aluminium-killed low-alloy steels susceptible to grain boundary AlN precipitations
and corner crack formation in cooling. The significant improvements achieved in
yield and quality with optimized thermal processing of the blooms are reported. Today nearly 100% of the CC production is hot charged thus providing a remarkable increase in production capacity and substantial energy savings.

KEYWORDS

Specialty steels; continuous casting; hot charging; in-line rolling; heat retention
tunnel; energy savings; yield; quality.

INTRODUCTION

The Aosta plant of DeltaCogne is a prime producer of long products in a wide range
of specialty steels from low-alloy construction steels to stainless, heat-resistant
and tool steels, as well as high speed and special purpose steels. The distribution
among the steel groups is shown in Table 1 for the 1987 production. Today about 65%
of the total production is continuously cast (CC) in form of one bloom size and near
ly 100% of all heats are being hot charged directly for subsequent roll cogging to
intermediate billet sizes (D-HCR process).

TABLE 1 1987 Production of Specialty Steels

Construction steels	47.2 %
Stainless steels	35.2 %
Valve steels	8.4 %
Tool steels	7.5 %
High speed and special steels	1.7 %

HISTORICAL BACKGROUND

When in 1975 the company decided to enter into continuous casting, the principal
production line consisted of a blast furnace, two oxygen LD converters and three
electric furnaces for a total yearly production capacity of about 350.000 t. The
innovative original concept of the CC facility with direct hot transfer of the cast
blooms to reheating and roll cogging was put in operation in 1977. A schematic layout of the major equipment is shown in Fig. 1. Despite   numerous initial operation problems this pioneer facility developed over the years into a successful production system providing remarkable improvements in yields, energy consumption and
operating costs. An 80 t heat of construction steel (or 50 t of stainless steel)
used to be cast into blooms of 180 x 240 mm on three strands in 60-70 min at strand
speeds of 1.0-1.1 m/min (or 40-50 min for stainless steels). Charging the blooms
into the reheating furnace took place with a track time of up to 60 min after torch
cutting due to slow lateral transfer (Fig. 1) and limitations in heating capacity
of the oil-fired three-zone walking beam furnace which was rated at 50 t/h max. for
hot charging under optimum conditions. The proportion of hot charged heats was high
in the early years but dropped with the increase in CC production because of sche-
duling/logistic problems and unsatisfactory surface conditions requiring bloom con
ditioning prior to rolling.

TABLE 2 Evolution of hot charged heats in
relation to CC production

	1985	1986	1987	1988*)	
Direct hot charged	93	96	98.5	99	%
CC Production	120	125	151	51	10^3 t
Construction Steels	54	46	55	37	%
Stainless Steels	46	54	45	63	%

*) 5 months

A continued effort to resolve the difficulties by modifying casting parameters and
paying close attention to the liquid steel quality through the introduction of metallurgical treatments in the ladle and better tundish designs resulted in improve
ments but a real breakthrough was lacking. In 1984 a major upgrading and moderniza
tion of the CC facility was initiated. The bloom size was increased to 200 x 275mm
to better accomodate the heats of 80/90 t from the UHP electric arc furnace. The
principal CC modifications completed in 1984/85 included the installation of a lad
le turret, tubular chrome-plated copper molds, magnetic stirring and level control

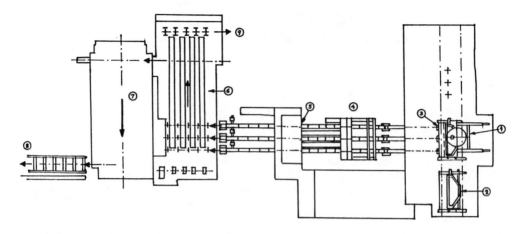

Fig. 1   Original layout of continuous casting-hot charging-roll cogging plant (1977).

1 - Ladle 50/80 t, 2 - Tundish 6 t, 3 - Molds 180 x 240 mm, 4 - Straightening, 5 - Torch cutting, 6 - Lateral transfer table, 7 - Oil-fired heating
furnace, 8 - To reversing mill, 9 - Air cooling of blooms.

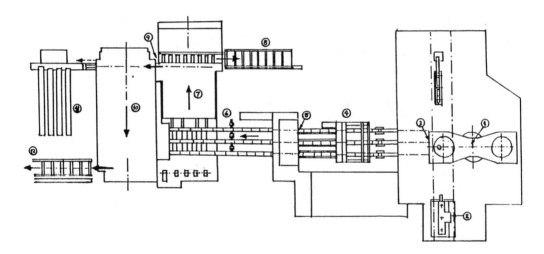

Fig. 2   Present layout of the CC-D-HCR faciliy (1987).

1 - Ladle turret, ladle 85 t, 2 - Tundish 12 t, 3 - Molds 200 x 275 mm,
4 - Straightening, improved design, 5 - Cutting torches more efficient,
6 - Runout table, 7 - Heat retention tunnel, 8 - Cold charging conveyor,
9 - Charging of reheat furnace, 10 - Gas fired reheat furnace, 11 - Air
cooling of cast blooms, 12 - To blooming mill.

in the molds, tundish weighing, primary and secondary cooling systems and new cutting torches. Significant improvements in bloom surface quality were achieved
through this extensive upgrading but the aluminium-killed low-alloy construction
steels still exhibited a high incidence of corner cracking resulting in reduced
billet yields from excessive conditioning. Metallurgical investigations and a lite
rature survey (1-3) indicated that the principal cause for this problem were inter
granular fracture associated with the precipitation of AlN in the proeutectoid
ferrite network. An optimization of the thermal regime of the blooms was initiated
in 1986 through controlled secondary cooling by air-water mist and a heat retention
tunnel over the lateral transfer table between caster and reheating surface, as
shown in Fig. 2. In 1987 the heat retention chamber was equipped with gas burners
for better and more uniform temperature control under varying production conditions.
With this optimized thermal handling system close to 100 % of the CC production is
now hot charged (Table 2), thus providing significant increases in production capa
city and yields at substantially reduced fuel consumption. The major technical data
of today's CC plant are listed in Table 3.

METALLURGICAL CONSIDERATIONS

Surface cracks in medium carbon and Al-killed low carbon alloy steels (carburizing
grades) cause serious problems in rolling of CC blooms. The carbon steels pass from
solidification to 600°C through three embrittling regions, as illustrated schemati
cally in Fig. 3 (1).

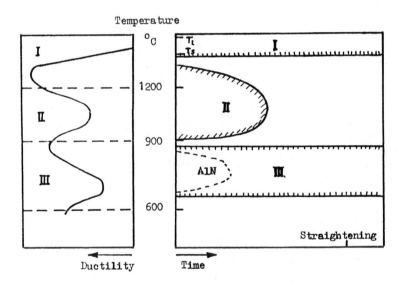

Fig. 3. Hot embrittlement regions in the solidification-
         cooling process of steel.

TABLE 3   Technical data of continuous casting equipment,
reheating furnace and blooming mill.

CC machine type	straight mold, strand bending and progressive straightening
Supplier	VOEST-ALPINE
Strand center distance	1800 mm
Mold size	200 x 275 mm
Corner radius	10 mm
Metallurgical length	15 m
Bend radius	8 m
Bending unit	progressive
Number of strands	3
Mold length	800 mm
Mold stirring equipment	Rotelec
Mold level control	Radioactive, Co 60, Berthold
Mold oscillation	Mechanical, lever type, $\pm$ 3 mm
Oscillation frequency	130 cpm at $V_c = $ 1m/min
Heat size	80/90 t
Ladle turret	rotating
Tundish size, shape, depth	12 t, T-shape, 800 mm
Tundish weight control	load cells
Casting speed	0.7 - 1.0 m/min
Primary cooling control	Based on Δ T, constant
Secondary cooling	Air-water mist, water flow rate regulated by casting speed
Torch cutting	Oxy-acetylene and powder
Bloom lengths	3500 to 7000 mm
Track time through tunnel	25 to 50 min.
Exit temperature from tunnel	840 - 880°C
Preheat furnace temperature	850°C
Reheating in walking beam furnace	1250°C (Construction steels)
Reheating capacity	80 t/h
Total reheating furnace capacity	160/170 t
Fuel consumption, average	$260 \cdot 10^3$ kcal/t
Reversing blooming mill	VOEST-ALPINE + DANIELI
D.C. Motor power, revolutions	1850 kW, 0-90-160 RPM
Roll diameter	700 mm
Manipulator	Sack ATR and RTR
Hot saw, blade diameter	1400 mm

Of primary concern are the zone II and III when the strand is drawn into the secondary cooling zone (II) and subsequently cools on the roller and transfer conveyors (III). Corners are more susceptible to cracking since they are not only subjected to quicker cooling but also due to larger unbending deformation stresses. Minute cracks between dendrite branches originating in region I during the very early phase of solidification of the shell (thermal stresses, phase transformations) may be filled with residual liquid rich in embrittling elements. With subsequent non-uniform cooling stresses these cracks may reopen in region II and III. More uniform cooling in the mold to produce an uniform shell thickness and improved secondary cooling control greatly help preventing the formation of solidification induced cracks (2).

The occurance of intergranular fractures in low-alloy aluminium grain-refined steel ingots and CC blooms has been known for some time. It was generally associated with the precipitation of aluminium nitride (AlN) in the proeutectoid ferrite network in region III. Increasing aluminium levels result in more proeutectoid ferrite and a coarser solidification structure with more microsegregation. Simultaneously, the precipitation of AlN is shifted to higher temperature because of the temperature dependence of the solubility product:

$$\log (Al) \times (N) = -6180/T + 0.73$$

It is, therefore, very important to control the aluminium content within a narrow range and to limit nitrogen to a reasonable level. The deleterious effect of alumi nium on the ductility of low-alloy steels in the temperature range of 400-900°C has been described by Ericson 1977 (3).

The other contributing event to poor hot ductility is the enrichment of residuals in the last solidifying areas between dendrites of ferrite and the accompanying peritectic reaction. This segregation of impurity (S, P, Cu) and alloying elements (C, Mn, Mo) to the remaining liquid can result in the formation of film-like low ductility phases. Alloys freezing completely to austenite have a low degree of microsegregation and are not prone to this cracking mechanism.

It can be concluded that maintaining the strand surface temperature near the top of region III will prevent the formation of precipitation induced cracks by AlN. The techniques of very soft and reduced secondary cooling for the purpose of al lowing the shell to reheat by the remaining latent heat of solidification from the core has been called the "auto-reheating method" and it is being practiced in slab casting (2). This principle is basically used to maintain elevated bloom surface temperatures in combination with a preheated transfer tunnel between torch cutting and heating furnace (4). Since the hot blooms enter the walking beam reheating fur nace with a higher heat content, it resulted in shorter reheating and soaking times (important for stainless steels), better heating efficiency, higher furnace through put and much enhanced operational flexibility for the interconnected units in the HCR line.

DESIGN OF HEAT RETENTION TUNNEL

In the initial concept of retaining a bloom surface temperature of ≥800°C, it was contemplated that the heat content of the cast blooms would be sufficient to exceed this temperature and at the same time obtain equalization and presoaking. The blooms have a surface temperature of 800-880°C and an internal temperature of about 1150°C after torch cutting. The well insulated chamber of 8 m length was designed to con tain 21 blooms of 200 x 275 x 7000 mm, weighing ca. 60 t. As shown in Fig. 2, the blooms are moved from the exit roller table in later direction to the entrance of the heat retention and equalization tunnel where a special device lifts the blooms onto the walking beam conveyor. The approximate time to pass the blooms through the tunnel was considered to be 30-60 min with an exit temperature of >800°C.

With the installation of the thermal optimization system in 1986, the existing
walking beam reheating furnace was modernized with gas burners for efficient heating from 800 to 1300°C and with two new recuperators to preheat the combustion air
to 650°C (previously 400°C). This latter change should decrease the natural gas
consumption by about 10%. Improved heat transfer by radiation is obtained too with
blooms in the equalized thermal condition. The initial operation of the heat reten
tion tunnel quickly showed that uniform bloom temperatures could not be obtained
under varying operating conditions. Starting from cold was particularly unsatisfac
tory but also between heats the chamber temperature dropped excessively. Therefore,
to attain a greater safety margin and to maintain temperatures consistently > 800°C
the chamber was equipped with gas burners to supply a total of 900.000 kcal/h. The
improvements in temperature uniformity and heat content led to the use of the
chamber as a holding unit with an increase in total heating capacity to 160/170 t
which is equal to two heats cast in sequence.

The construction of the preheating chamber is in general based on modern walking
beam furnace technology with complete refractory hearth lining and light weight
fiber wall and roof linings. Six gas burners are mounted on the side walls whereby
the burners can be operated individually according to the real heating needs. The
large front door presents a major heat loss area that can only be compensated with
the two burners near the opening. An additional two burners are frequently used to
maintain an adequate chamber temperature between heats while the remaining burners
are usually on standby for start-up or to boost the preheating temperature. The
burners are operated by semi-automatic control based on thermocouples mounted at
the entry and exit end of the chamber to monitor the temperature.

                    OPERATING AND QUALITY/YIELD RESULTS

Before the installation of the heat retention tunnel the surface temperatures of
the blooms upon charging into the reheating furnace varied widely, as illustrated
in Fig. 4 on a typical sequence cast of two heats of construction steel. The irregular charging conditions caused by delays and furnace heating capacity limitations
required many restrictions in metallurgical operating practices to minimize losses
resulting from cracks by AlN formation in the temperature range 800 to 600°C.

   lacing the heat retention chamber in front of the reheating furnace improved
the thermal condition to some degree and provided a desirable equalization of the
bloom temperature at a higher caloric value. This situation with a chamber tempera
ture of about 250°C initially is shown in Fig. 5 for two heats passing through in
sequence. The slow recovery of the bloom surface temperatures was not satisfactory;
single heats showed a thermal behavior still worse. Even though grinding losses
were reduced (see Fig. 7), the thermal characteristics of the tunnel were not acceptable for a consistent improvement from the first to the last bloom and it was
quite clear that heating of the chamber was unavoidable for optimum operational
and quality results. After the installation of the burners, the situation improved
drastically under all operating conditions, as shown for a single heat in Fig. 6
where bloom surface temperatures are uniform and always over 800°C. The exit temperatures from the heated tunnel are consistently higher than the charging tempera
tures. This modification has provided the necessary flexibility in the D-HRC
operation and all restrictive practices could be eliminated for aluminium-killed

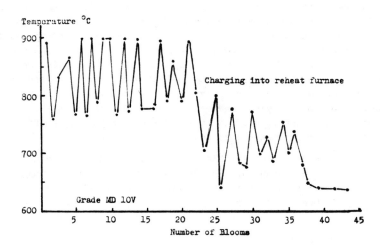

Fig. 4   Bloom surface temperatures at entrance of reheating
         furnace before installation of heat retention tunnel.

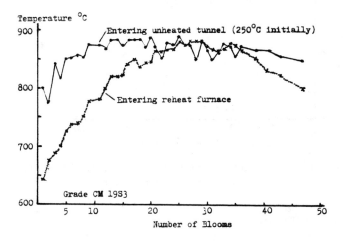

Fig. 5   Bloom surface temperatures at charging into unheated
         heat retention tunnel and at the reheating furnace.

fine grain construction steels. Significant benefits were also obtained in stainless steel heating with this change.

The extensive upgrading of the CC facility and the advances made in liquid metal

treatment in the ladle together with the final optimization of the thermal handling
resulted in largely defect-free cast blooms with excellent surfaces. Direct hot
charging for rolling became a reality without practice restrictions since intermediate bloom conditioning could be eliminated with very few exceptions. The remark
able decrease in billet conditioning losses on construction steels over the past
four years is shown in Fig. 7 which also lists the major corrective actions taken
in this period. The benefits of using the preheated chamber as an equalization and
preheating unit for two types of stainless steels are well illustrated in Fig. 8
showing considerable billet yield increases. Advanced composition balancing for
better hot workability contributed to this success with the new heating practice.
The general quality improvements obtained with CC was reported by Gubiotti 1982 (5).

Monitoring the fuel consumption for bloom reheating yielded the results summarized
in Fig. 9. The achieved savings are substantial and in line with projected reductions. Further work in this area is under way for additional optimization in heating efficiency and lower operating costs.

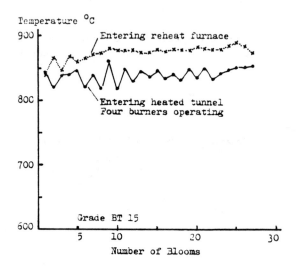

Fig. 6   Bloom surface temperatures after installation
of burners on heat retention tunnel.

FUTURE DEVELOPMENTS

The continuous casting-HCR plant has reached a high operational standard producing
reliably high yields and excellent surface and internal quality in blooms of special
ty steels. The next developments involve continued refinements of the operation
practices and the successive introduction of automation in specific areas to impro
ve further product consistency and reduce costs. A first project is dedicated to
the areas "heating of blooms" and "roll cogging-billet cooling" with the intention
of optimizing the thermal regime in total, improve roll pass design, billet cutting
to length and controlled cooling. The second level of automation concerns the CC
machine, the coordination of critical casting parameters, closed loop regulation

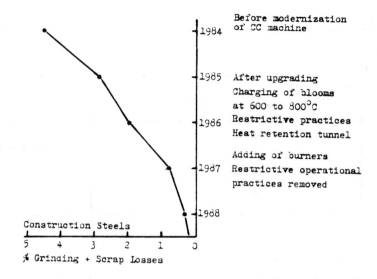

Fig. 7   Decrease of billet conditioning and scrap losses on
         construction steels caused by bloom surface cracks
         in rolling.

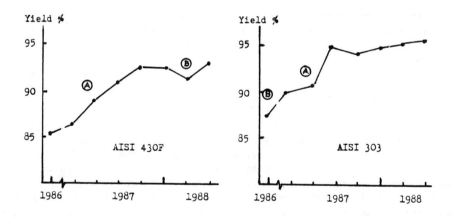

Fig. 8   Increases in billet yields after conditioning of two
         stainless steels produced by the new heating practice
         (A) and through advanced compositional balancing (B).

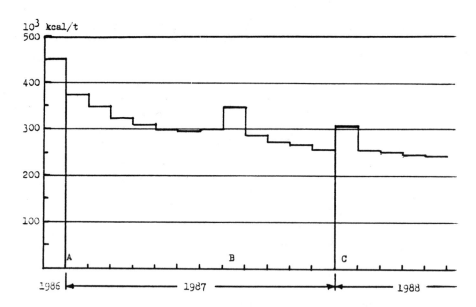

Fig. 9   Reduction of specific heat consumption in kcal/t for heating
         of blooms with improved thermal handling.

         A - Installation of heat retention tunnel
         B - Shut down; installation of burners on tunnel
         C - End of year shut down

and a process data acquisition and evaluation system. In line with these projects
are also efforts to find an advanced hot billet inspection system to detect defects
immediately during processing to shorten the production cycle and accelerate corrective actions.

CONCLUSIONS

The experience and the results acquired at DeltaCogne on hot charging and in-line
rolling of CC blooms confirm fully the expected savings in energy consumption, the
cost reductions in material handling, and the improvements in production capacity
and operational flexibility. Coupling of several production units in line causes
generally logistic problems when a wide variety of materials must be processed. The
installation of the heated tunnel which also serves as an equalization, preheating
and holding chamber has greatly alleviated these scheduling difficulties. Of equal
importance are the simultaneous achievements in improved product quality, consisten
cy and in higher billet yields. Direct hot charging and hot transformation of CC
blooms and billets will undoubtedly be employed more extensively in the future as
the most cost-effective processing route for high quality steel products.

REFERENCES

(1) Trans. ISIJ, 25 (1985), No. 7 and 8, Special Issues for the 70th Anniversary.

(2) Tsubakihara, O., Trans. ISIJ, 27 (1987), 82 and 86-87.

(3) Ericson, L., Scand. J. Metallurgy 6 (1977), 116-124.

(4) Schlatter, R., Internal Communication 24/10/1985.

(5) Gubiotti, R., G. Melotti, R. Medori, Presentation at Associazione Italiana di Metallurgia, Milano, June 10, 1982.

KEYWORD SUBJECT INDEX

AUTHOR INDEX